The Puzzle Solver

The Puzzle Solver

A SCIENTIST'S DESPERATE QUEST TO CURE THE ILLNESS THAT STOLE HIS SON

Tracie White with
Ronald W. Davis, PhD

NEW YORK

Hachette Books
Hachette Book Group
1290 Avenue of the Americas
New York, NY 10104
HachetteBooks.com
Twitter.com/HachetteBooks
Instagram.com/HachetteBooks

First Edition: January 2021

Published by Hachette Books, an imprint of Perseus Books, LLC, a subsidiary of Hachette Book Group, Inc. The Hachette Books name and logo is a trademark of the Hachette Book Group.

Library of Congress Cataloging-in-Publication Data has been applied for.

ISBNs: 978-0-316-49250-8 (hardcover); 978-0-316-49249-2 (ebook)

Printed in the United States of America

LSC-C

Printing 1, 2020

For Whitney

Contents

The Puzzle Solver

Introduction

Everyone knows someone with a mysterious illness that goes unmentioned and gets ignored. Often, it's a chronic disease with no known cause and few treatments. The illness simply refuses to dissipate. As years pass, doctors grow frustrated; families grow weary. The sick grow hopeless. A curtain falls, to hide the unimaginable and enduring pain. And silence remains. This is an attempt to lift the curtain from over those sick people who go unseen, untreated, and forgotten. But this book will also bring those sick, and the ones who love them, hope.

Whitney Dafoe, a young fine arts photographer who traveled the world, now has one of those ongoing illnesses that often gets ignored. Once he was healthy and strong, but an illness crept up on him slowly, incrementally taking away bits of his life. First it took away his energy, then his ability to walk any farther than the few steps to his bathroom. Eventually he could no longer talk or eat. He couldn't leave his room or even bear to have others join him there. Symptoms of severe fatigue, gastrointestinal problems, muscle pain, and neurological disorders plagued him for years before he even got a diagnosis. He was told there was nothing wrong with him, that he was crazy, that if he exercised he'd be fine. Finally, eight years ago, he got a diagnosis. He has chronic fatigue syndrome, now known as ME/CFS, a disease with no known cause, no approved treatments, and no cure. And still, it strikes twenty million people worldwide.

Yet the course of this story has changed due to Whitney's father, a legendary Stanford University scientist. Ron Davis, whose discoveries helped launch the Human Genome Project, is now on the hunt for the prey that took his son away. Beginning with a drop of his own son's blood, he launched a search for the cause and, from there, a treatment and a potential cure for this often forgotten and stigmatized disease.

ME/CFS, like many other illnesses, usually strikes healthy people out of the blue, changing their lives forever. It brings fear and uncertainty, like AIDS or the coronavirus. Most go undiagnosed. One quarter get severely ill like Whitney, bedridden or unable to leave their homes. Many lose their jobs, and even family and friends. Without a definitive lab test to diagnose the disease, they are often disbelieved by doctors and face obstacles getting medical insurance or disability. No one knows just how many end up homeless.

Despite a new acceptance within the scientific community as a biological disease, the illness is still little known with not much funding for research. Patients are still mocked for being nothing more than tired, or lazy, or crazy. Once called the "yuppie disease," it remains hidden away, with a long and winding, messy history that also must be unveiled.

Ron Davis has used his prestige within the scientific community to speak out, share his new research discoveries on this disease, and tell his son's story, the same story that millions of others have suffered and continue to suffer. It's a story that needs to be told.

Chapter 1

The House

IN DECEMBER 2016, SOMETIME in the middle of the night, Whitney Dafoe slowly used Scrabble tiles to spell out this:

"Can't take care of myself. Don't know what to do."

This was followed with five more tiles: D...Y...I...N...G.

Then he pushed a button next to his bed that rang a bell throughout the house to summon his parents, who were caring for him around the clock, to his room.

Later, when I asked his father, Ron Davis, a legendary scientist, how he responded when he saw what his son had spelled out in tiles, he simply said, "I did not answer him—he cannot tolerate anyone speaking to him."

That week, Whitney was taken in an ambulance to the hospital to get a feeding tube inserted into his abdomen. He needed it to stay alive.

Two months later, I drove up to Ron Davis's home in Palo Alto, California, for the first time on a chilly winter's evening. A gentle breeze touched my cheek as I climbed out of my old, green Honda Civic, but my heart felt heavy. I had come to interview Ron and his wife, Janet Dafoe, to hear Whitney's story. (Whitney's parents, I would also learn, chose to share their last names between their two children. Whitney has his mother's last name, and their daughter, Ashley, was given her dad's.) Over several decades as a journalist, I had gained a reputation as the go-to writer for tragic stories, so it was natural my editor at *Stanford Medicine* magazine had assigned me this story about a father trying to save his severely ill son.

I'd written articles for the magazine before about amputees who survived the 2010 Haitian earthquake, Cambodian refugees with PTSD, the deplorable lack of health care on Native American reservations, and myriad other science stories about horrific medical conditions and the search for cures. I was proud to be able to give a voice to these often neglected stories, but I also had a personal motivation. I'd been devastated myself decades before by a late-term miscarriage. The grief eventually lifted, but it left me with a chronic case of insomnia that I struggle with still. Along with that came a growing obsession with writing stories about people living with chronic conditions. I wrote about cancer and Alzheimer's disease, clinical depression and foster children living with PTSD. I'd look into people's eyes when I interviewed them, searching for answers. How did they do it? How did they carry on?

Still, for some reason, when confronted with this story of the daily pain and suffering of a young man imprisoned in his own body with his grief-stricken family standing at his side, it unnerved me. Perhaps it was the years of his total and complete isolation that scared me most or the fact that his photos looked too much like my own grown son. Whitney couldn't eat or speak. He was completely voiceless. Now thirty-one years old, he lived twenty-four hours a day, every day, in the back bedroom of his parents' home. I'd talked briefly over the phone with both Ron and Janet when arranging the interview. They told me Whitney had been diagnosed with chronic fatigue syndrome more than five years before, but that for the past two years, his condition had deteriorated into a coma-like state. When I asked if I could interview him, they'd said no. His mother was the first to say it. Whitney was far too sick for anything like that.

"I understand," I told her. Though at the time I didn't fully.

At the time, I was working for Stanford medical school in an office building that happened to be across the street from Ron's science lab, the Stanford Genome Technology Center, in Silicon Valley. As a science writer who often wrote about faculty research, I was aware of his reputation. I'd never met him before, but when my colleagues talked about him, it was with awe and respect. Some rumored he could be in line for a Nobel Prize one day. So it was big news that Ron Davis, PhD, had suddenly changed the course of his genomics research to study this little known disease and find a cure for his son.

I had read through the smattering of media stories about Whitney and Ron online. In a 2015 *Washington Post* article titled "With His Son Terribly Ill, a Top Scientist Takes On Chronic Fatigue Syndrome," Ron Davis, the story said, developed techniques for gene mapping that were later used in the Human Genome Project. In 2013, the *Atlantic* ranked him on a shortlist of the world's greatest living inventors alongside SpaceX founder Elon Musk and Amazon founder Jeff Bezos.

His son, on the other hand, had been an artist. Whitney used to be an award-winning photographer who traveled around the world. He had helped to build a nunnery in the Himalayan mountains, ridden a motorcycle across India, and visited all fifty American states. Now he was fed intravenously and could barely tolerate sounds. His parents, and the medical personnel who saw him, wore plain clothing when they entered his room because bright colors or any patterns could make him feel worse. I lingered over the before and after photos of Whitney that ran with the story. One showed a smiling Whitney in a red plaid jacket, tall and handsome, with bushy brown eyebrows and kind eyes, posing in front of the blue waters of the San Francisco Bay. In the next, an emaciated body lay on its side in bed. Ron knelt there, at Whitney's side, shaving his son's head because he had grown too weak to wash his own hair.

Standing in this family's driveway, as a parent of two children myself and having many years ago lost a third, I just wanted to turn my head and drive away. The suffering of a child is so hard to bear. Instead, I chastised myself for being cowardly, took a deep breath, and turned to walk down the tree-lined driveway, pausing occasionally to stare up at the stately old house with columns on the front porch and wisteria climbing its walls. Somewhat hopefully, the house was decorated with twinkling white lights strung across a wide, green lawn and colorful Tibetan prayer flags hanging from a circular front porch. I would learn that back when it was first built in 1889 in the upscale neighborhood now known as Professorville, this house had a palm-lined circular driveway that curved past tennis courts, a coachman's quarters, and an outside kitchen, a sign of old Palo Alto wealth. When Ron and Janet moved into the house in 1983, their son just an infant, the tennis courts had long ago disappeared, and the aura of old wealth had changed to a quaint nostalgia.

I walked across the front porch and paused to read a handwritten note stuck to the front door: "Please do not knock or ring a bell before 3 p.m. Call or text. Very sick person."

Stupidly, I checked my watch, even though it was obviously well past 3 in the afternoon. I knocked softly. Janet opened the door and invited me inside.

Like many people, I knew little about chronic fatigue syndrome when I arrived at the house that evening. Reading through the *Post* story had jogged my memory about an old article I had written on CFS years earlier, back when I was working as a reporter for my hometown paper, the *Santa Cruz Sentinel*. It was a mysterious disease with no known cause or cure, often lumped together with other so-called contested diseases—like chronic Lyme disease, fibromyalgia, and Gulf War Syndrome. I'd heard it was a "woman's" disease, which often meant, insultingly, psychological in origin. Therefore forgotten and tossed aside. I had no idea patients with CFS could get as sick as Whitney. Many people, including physicians, didn't even believe it was a real disease. Some said these chronically fatigued "sick" people were just suffering from depression or were "malingerers" who wanted to collect monthly disability checks. I knew there was an active CFS patient community that protested such propaganda and cited science to back them up. I would hear more about it from Whitney's parents that night.

I followed Janet into a large foyer, and she took me on a quick tour of their home. It was a house overflowing with a lifetime's worth of science books, paintings, and Whitney's photographs. The detritus from years of raising children filled every nook and cranny with baby photos, Whitney in Little League, and Ashley in ballet. I slowly began to relax. It reminded me of my grandmother's home in Salt Lake City, with crystal doorknobs and those old-fashioned light fixtures that you had to push hard into the wall to turn on. I felt comfortable in Whitney's home. Janet opened a door to the basement revealing a creaky-looking staircase just like my grandma's, where my cousins and I used to play. Basements are a rarity in California. I wondered if Janet canned peaches like Grandma Deedee had.

Janet had short gray hair like my Grandma DeeDee. She was a child psychologist with a PhD from Stanford.

And, at least that day, she had her guard up. She doesn't trust me yet, I thought. We walked back through the foyer, then settled in the library, on comfy couches with shelves full of old books, a television, and their pet tortoise, Rex. This was obviously where the couple escaped to when they had a chance to relax.

"Ron feels a huge amount of responsibility and stress," Janet said, as she sat down on the couch. I pulled out my reporter's notepad. Ron joined us in the room and quietly sat by her side. I introduced myself, and we shook hands. He was wearing old jeans with a cowboy belt buckle and sneakers, his brown cowboy hat on the coffee table in front of him. Not a whiff of arrogance came off this man. He had a gray beard, glasses, and a kindly demeanor that lent him more of a grandfatherly air than any kind of snooty professor with long lists of big science awards. Mostly, he just looked tired. At seventy-one, Ron had undergone an aortic heart valve replacement two months earlier but returned to his CFS research as quickly as he could, Janet said.

"He also has to keep all his other grants up," she told me. "He spends all the waking hours he can thinking about CFS research."

Ron smiled sadly, then added, "It is enormous pressure. We have to figure this out very quickly because millions of people are suffering and my son is dying."

I didn't know how to respond. I tried not to think of how it might feel to have your grown child so sick, nearing death. Looking around the living room, I could see how well they'd seen to it that he and Ashley had a safe and grounded childhood. The thought of losing him to an illness with no cure and one that they did not have any control over—I felt a great sympathy for Ron especially. To feel the responsibility of finding a cure not only for his son but for so many other people's loved ones who depended on him must've been a huge burden for this gray-haired man. My heart hurt for them as Janet began to tell me more about Whitney's story. As she spoke, I had the feeling she had told this story many times and would tell it many more. It was as if it was her job as a mother was to make the world both understand and believe in her son's story.

Maybe it first started with a case of mononucleosis in high school, Janet told me. They didn't know for sure. Or it could have been at the age of twenty-one, after a trip to Jamaica during college when Whitney experienced dizziness, a severe headache, and nausea—symptoms that never seemed to completely abate. More signs of illness came two years later, when he traveled to India for the better part of a year.

Whenever he could, Ron veered the conversation back to the science. Talking about his son's sickness was obviously painful for him, but he understood science. It made much more sense to him. He explained how the disease affected multiple systems of the body including the immune, nervous, and metabolic systems. He told me he had been part of the Institute of Medicine (now the

National Academy of Medicine) committee that the year before had published a report on CFS. It listed symptoms that range from chronic pain to cognitive problems, brain fog, gastrointestinal issues, sleep disturbances, and extreme fatigue. The flagstone of the disease is a symptom known as post-exertional malaise, PEM, which means symptoms worsen after any kind of physical, mental, or emotional exertion—sometimes, in severe cases, it can be something as simple as brushing your teeth. Also, sleep doesn't help the fatigue, and sleep cycles go haywire.

"So many people will ask us, 'Has your son tried caffeine?'" Ron continued, shaking his head. "They think CFS just means you're really tired. Caffeine does not help with this disease."

At the sound of a faint bell, he suddenly stopped talking and stood up.

"That's Whitney," Janet told me. "He's ready for his dad." Whitney usually woke up in the late afternoon or early evening, Janet explained. Ron's daily routine was to care for Whitney then, giving Janet, who stays up most nights with Whitney, a break. He'd clean up the room, pick up tissues, fill containers with water. It was his job to prepare the IV line that delivered the life-sustaining TPN, liquid nutrients, along with medications and supplements. The feeding tube inserted into Whitney's abdomen a few months earlier was beginning to help add more nutrition, but each additional supplement still had to be tested carefully.

"When I walk in the room, he's still as death," Ron told me before he left. "If his feet are left uncovered, it means he wants me to change his socks." Janet explained how it took awhile for Whitney to prepare mentally for someone to enter his room to prevent him from "crashing." I didn't know what this meant, but it sounded bad. She said they would sit in the hallway outside his door and peer through the keyhole to watch for Whitney's signal that it was OK to enter. This was hard to imagine, so I asked if I could see how it worked. I followed Ron to the hallway that led to Whitney's room; then he signaled for me to watch from there.

Ron didn't enter Whitney's bedroom right away. Instead he sat on a chair outside the door and peeked through a keyhole waiting for his son to find the strength to sit up and pull a blanket over his shoulders, signaling it was OK to come inside. I could see it was one of those old-fashioned keyholes big enough to see through.

"He can hear me sitting outside," Ron explained to me later. "It can take him a long time to prepare. Sometimes I wait for hours at the door. I once watched him take thirty minutes to move his cap from his lap to his head."

After some time had passed, I left Ron waiting there and returned to the library to join Janet on the couch. It was getting late, and I had a long drive back home to Santa Cruz that night. As I stood to go, Janet pointed to a book on one of the shelves titled *Osler's Web*, which had been written decades ago by Hillary Johnson, an investigative reporter who had CFS herself. "The folks in the patient community think of it as their Bible," she said. "It tells the story of how the medical establishment failed this disease."

She told me that we should take another look at the back of the house before I left. When I entered the kitchen, I realized that it had basically been transformed into a hospital ward for Whitney. The large, wide-open space was filled with boxes of syringes and medicine bottles. Just off the kitchen, a closed door led down a short hallway to Whitney's room. Janet put her finger to her lips telling me to shush and softly opened the door.

I followed her down the hallway, and ever so gently she opened the door to Whitney's bedroom. From the hallway I caught a glimpse of him from behind. Ron was on his knees at his side fiddling with the IV line. Whitney didn't see me. He was sitting up hunched over in bed, his bare spine a long string of white pebbles dripping down from thin neck to slender hips. That was in the spring of 2016, when his six-foot-three-inch frame weighed just 140 pounds, and he was starving to death. I wouldn't see him again for two years.

The magazine article I was writing, "The Puzzle Solver," was published that spring, but that was only the beginning of the story. Even after the article, I'd think about Ron's words and Whitney's years of complete solitude. I'd go home at night and hug my son, who was now at an age when he was thinking about his career and starting real life. I couldn't imagine being in Janet's place, but sometimes I thought about it. I wanted to do something to help. I needed to talk to more people who had the disease. Few in the medical community had taken it seriously, and there weren't that many places to look. I needed to know if they were experiencing what I had seen in Whitney's room. When I saw Whitney, I knew there was no way he could just be depressed, as some had suggested. He was suffering, and it was beyond mental anguish. Something was eating him alive.

I reached out to Laura Hillenbrand. In addition to being a CFS survivor, she is the best-selling author of *Seabiscuit* and *Unbroken*. She had written about her struggles with a similar severe case of CFS in a 2003 *New Yorker* article, "A Sudden Illness," which I remembered reading. She had also written entire books while sick in bed. I emailed her publisher and received a response back quickly that she'd be willing to talk to me.

She was a nineteen-year-old student at Kenyon College in Ohio when she first got sick. One morning she woke up and couldn't sit up in bed. There were chills, swollen lymph nodes, dizziness. The campus physician ran test after test but couldn't find anything. As time passed, she grew too weak to continue her studies, dropped out of college, and moved home to Maryland to live with her mother. She lost twenty pounds. Her lymph nodes stayed painfully swollen. At night, she had chills, and the ground swayed.

"Sometimes I'd look at words or pictures but see only meaningless shapes. I'd stare at clocks and not understand what the positions of the hands meant.... Words from different parts of the page appeared to be grouped together in bizarre sentences: 'Endangered Condors Charge in Shotgun Killing,'" she wrote in the article.

A doctor said the problem was in her mind and sent her to a psychiatrist. The psychiatrist told her to see another doctor. She saw more doctors. Medical tests all came back negative. Finally, a doctor diagnosed something called Epstein-Barr virus syndrome. I didn't yet know what type of virus this was, but viruses, I knew, caused infectious diseases.

He gave her supplements, but they didn't help. Finally, she began seeing different specialists. After a lengthy exam with an infectious disease specialist at Johns Hopkins, she was diagnosed with CFS. He too suspected that it was viral in origin. She was in her thirties when she wrote the article.

It had been more than a decade after the article was published, and I didn't know how she'd feel talking about it when I called. All that pain and suffering, she told me, has brought with it at least one gift of which she's proud.

"The experience of suffering has brought me great compassion for the suffering of others," she said.

Then, matter-of-factly she described to me what it felt like to be trapped in bed with severe CFS.

"The exhaustion is so profound, it's a struggle to breathe. It's a struggle just to lie there. Every effort goes into just staying alive." At first, she was angry at family and friends who had abandoned her. Angry at doctors who told her she was crazy. "It was arrogance paired with ignorance. The easiest way to deal with someone with this disease is to deposit them in the waste bin of psychiatric disorders."

I thought of Whitney lying in bed. How had she survived all this?, I wanted—I *needed* to know. I asked her about suicide, and she said, yes, she had thought about it. But somehow she got better enough so that she could begin to form sentences in her mind and write stories from bed. It was excruciatingly slow work; CFS made everything she did slow. And so she chose to write about fast things. She used to be an athlete, and she missed motion so badly. So she wrote about a racehorse and an Olympic runner. And then, after being unable to leave Washington, DC, for more than twenty-five years, she very slowly began to make plans to move. It was a long, arduous path, she said, that led to her eventual limited recovery, a combination of new medications, constant restraint and patience, tiny steps forward over years. When I talked to her, she had just made a cross-country car trip from the East Coast to a new home in Oregon with her boyfriend. She prepared two years for that trip to deal with her vertigo, taking practice car trips beginning with five minutes, then adding a minute or two. Every trip, she said, was a whirling nightmare, but it worked.

"We drove 4,300 miles," she told me, the amazement still lingering in her voice. "Through the Rockies and the Badlands all the way to Oregon."

"That's so wonderful" was all I said, but my mind went right back to Whitney.

The phone call had made me hopeful. If Hillenbrand could feel better, have a relationship, and travel again, so could Whitney, right? I had a number I had read in my mind: twenty million people worldwide suffering from CFS. If they could be cured, that would change so many lives, but there was so little known about a cure.

Bits and pieces about this mysterious disease continued to nag at me as time passed. I had read that a chronic fatigue syndrome "epidemic" had first started in Incline Village, a small tourist town on the shores of Lake Tahoe in the 1980s. That was just a short four-hour drive to the ski slopes for me, a trip I had done many times as a college student at Berkeley. Maybe there was a trail to unraveling Whitney's illness, one that unknowingly started there, could circle back to Ron's laboratory in the building just across the street from my office, and end with Whitney and whatever story was hiding with him inside that bedroom. If Hillenbrand got better, then there was hope for Whitney, wasn't there?

Chapter 2

The Invisible Patient

Two years later

THICK, DARK LINES. THAT was all I saw at first. I pressed my eye to the keyhole just below the brass doorknob that opened to Whitney Dafoe's bedroom, where he had been sequestered for five years. I leaned back in the folding chair and closed my eyes. The dark lines slowly came into focus, just above a pair of deep, blue eyes—eyes like his father's. Ashley had always been jealous of her brother's long, black lashes, and I understood why. They were lovely. Those eyelashes, Whitney's eyes, were blocking my view into his room, I realized. I was that close to him.

It had been two years since I first visited Whitney's family. During that time, he had remained too sick for me to interview him, but still I found myself driving my Honda back to the old house in Professorville, trying to better understand the mystery of the invisible patient living there. I watched repeatedly as both Ron and Janet, sometimes Ashley, sat outside his door, peering through the keyhole, waiting for Whitney to signal it was OK to enter. Now it was my turn.

Janet had scheduled a surgical procedure for Whitney to have a new feeding tube inserted to replace his old one on this day in December. Over the two years since that first feeding tube was inserted, the plastic tubes kept breaking, and Whitney would need to make an emergency trip back to the hospital to get a new one inserted. The tubes just aren't built sturdy enough to last for years, and without them he could die. So Janet scheduled the trips to replace the feeding tubes before they broke once every three months. This was the only time Whitney ever left his room.

As the day's organizer, Janet was hypervigilant, constantly running interference between the timing of the ambulance drivers, the surgery appointment, and preparing Whitney for the day without making him suffer more than he already had to. While Janet was busy on the phone with the ambulance service, I walked over to join Ron in the kitchen. He wasn't feeling well when I had first arrived, and was lying on his back on the floor. He had vertigo and was dizzy and nauseous, his gray hair ruffled. "I'm nervous," I finally whispered to Ron.

"Me too," he whispered back, with a smile.

I'd entered that nebulous area, where the reporter gets too close to her subjects. I had become somewhat of a fixture in the old house, as the designated note taker to document this journey, constantly snooping around for clues in an ongoing obsession to learn as much as possible about Whitney. The family had grown used to me, maybe even had begun to like me, but I had never been in the house to witness these hectic trips to the hospital. I hoped that

I might be able to see Whitney up close, maybe even meet him. Maybe not. Janet warned these ambulance trips were difficult for Whitney, and they never knew how much stimulus he could handle.

Any disruption to his schedule could make him crash, unable to think or move. Janet had all these balls in the air trying to get Whitney to the hospital, and I was trying not to be the extra one to make them all come tumbling down. Ashley would arrive shortly to help Ron clean Whitney's room and change the bottom sheet on his bed after he left. My role that morning was to sit at the end of the hallway connecting Whitney's bedroom to the kitchen until Ashley arrived. And not to be too quiet. Whitney would hear that someone was sitting outside his door and know he needed to prepare for the day. It's a gentle way of waking him. He wouldn't know it was me, of course, because he still didn't know anything about me.

I took another deep breath and bent my head once more to the keyhole, feeling uncomfortably like a voyeur. But I was anxious to see more of Whitney than just a glimpse of him from the back.

A new image. A thin green string stretched across white skin. A necklace. A neck. Whitney was awake and moving. Oh my gosh, I almost gasped. I sat back up, took another breath, and leaned over again. Click. My breath came faster now. A patch of brown hair on the back of a head. The young man who had disappeared behind this door so many years ago had grown older. His hair was thinning now, perhaps from aging or maybe just from years of lying still on his back, his head resting flat on a pillow.

Piece by piece, image by image the invisible patient began to appear. In fact, he wasn't invisible at all. He was flesh and blood, water and bone, blood and cells and DNA. He had white skin and black lashes and wore a string necklace. He didn't eat; a feeding tube kept him alive. He rarely slept. He couldn't talk. Sound physically harmed him.

As the morning moved along, and the sun rose higher in the sky, I was replaced in my post at Whitney's door by Janet. Ron, now recovered from his vertigo, was steadying himself by leaning against the island in the middle of their farmhouse-style kitchen to prepare his son's medication, and I was watching him. Carefully, wire-rimmed glasses tipped low on his nose, he filled three syringes with liquid Ativan, an anti-anxiety medication. *Tap, tap*—he tapped a fingernail against the last plastic tube waiting for a single round bubble to float to the top of the liquid and pop. This was a prescribed drug for Whitney to relieve the added stress in preparation for his ambulance ride. Like many others with ME/CFS, Whitney experienced sensory processing disorders, and that was what made it difficult for him to tolerate sound, vibration, or touch. Just moving him from his bed was painful.

Ron looked up as Janet entered the kitchen; she patted him on the shoulder and took the three filled syringes. I heard her shuffle down the hall to Whitney's bedroom.

The couple, married almost fifty years, often move in tandem like this. One picking up where the other leaves off. A team.

Ashley arrived dressed in black yoga pants, her long blonde hair pulled back in a ponytail. With the grace of the ballet dancer she once was, she too crossed through the kitchen to the closed hallway door that leads to Whitney's room. Her tiny, white fluff-ball of a dog Frankie was tucked under her arm and started to yap. Janet's head popped out suddenly from behind the hallway door. She grinned and reached for the dog. She too was dressed in black, as were Ron and I because of Whitney's sensitivity to colors. That was our uniform for the day. Janet squeezed Frankie tight, laughing as the dog's long pink tongue washed her face. Ashley set Frankie down on the kitchen floor, and together they walked back to Whitney's bedroom, with Ron following. Then Janet waved to me to follow.

"Stay outside of the bedroom in the hallway," she instructed me. She'd told me earlier they'd discovered that Ativan did more than calm Whitney's nerves. It also seemed to calm some of his sensory processing difficulties, making sounds and movement more bearable. When he took the drug, he could communicate with his eyes, and pantomime using his hands and arms and facial expressions.

I watched carefully from the door as the family, huddled together in Whitney's small bedroom, gathered around his bed as if in prayer. These moments, I knew, were so rare, they were almost sacred, and emotions ran raw. Ashley sat cross-legged on the floor in front of her brother's bed, her eyes rimmed in red from tears. Ron, smiling softly, reclined on the carpeted floor just behind her, his long legs stretched out straight. His gaze drifted outside to the backyard. The winter sun sat on the leaves of an oak tree, painting the room's carpet with reds and gold. Janet stood next to the bed, beaming down at her beloved boy, waiting for him to acknowledge them. After the drug worked its way into his blood system, Whitney emerged from his comatose-like state. His eyes opened. The family had told me that usually, at this point, he would look up at his dad with his eyebrows raised, asking, Is there a cure yet? His father would shake his head no, then make fist pumps with his hands, meaning he's working hard. And then Whitney would begin to pump his fists too, asking for his dad to work harder. Then they joined together, fists pumping out like boxers. A few hours later, as the drug began to wear off, Whitney's newfound energy would slip away. In tears, he would head back to the abyss. Ron, Janet, and Ashley all would begin to sob. And then, suddenly, he was gone, alone again in his comatose-like state.

But today was different. He raised his hands to his face and clicked an invisible camera. Ashley sprinted from the room to get a real camera, her high-end Nikon. I couldn't believe what I was witnessing. I had only expected to see Whitney alert for a few minutes; now he was sitting smiling and interacting with his family. Tears began to fill my eyes. He was still unable to speak, but through pantomime he told them that he wanted a photograph of the family together. He could not hold the camera, but from his bed with two ice packs covering his concave stomach, he gave instructions to Ashley on just how to stage each photograph. His hands motioned his parents to move in together for a shot. The professional photographer still knew how to command a room. I could feel a hope fill the room. Ashley even laughed.

Early on, before he got too sick, Whitney had begun to document his illness with plans of someday making his own film about the disease. Now it appeared he still wanted to tell his own story. Maybe I could help him.

When the ambulance crew arrived, I left my hallway post and walked out the kitchen door to go around the backyard, past the pool, to a small patio just outside Whitney's back bedroom door, where he'd exit on a gurney.

I waited there, the silence deafening. A squirrel crashed through the redwood trees. A black crow cawed once, then twice. The thud of a squirrel landing shuddered up through the ground. Two EMTs appeared, exiting backward through Whitney's door, carrying him on the gurney. He would be taken to El Camino Hospital in nearby Mountain View, unit 2B, an isolation room usually reserved for patients with contagious diseases. It would be quiet there. He was wrapped in two soft brown blankets and wearing noise-blocking headphones. Suddenly Whitney motioned for them to pause as his eyes lifted heavenward to stare up at the wide, blue sky, his mouth dropping open with awe. It took my breath away. I watched as the crisp winter air touched his cheek. The sun. The breeze. Then the wheels of the gurney started to roll, squelching over wet, soft leaves. The air hung damp, washed clean by a recent rain.

The gurney rolled past the garden where Whitney once loved to plant flowers. African daisies decorated the walkway. He caught my eye as he rolled past, furrowing his brow in curiosity at the stranger standing there. It was a relief to be seen. Then he disappeared around the corner, and I followed. As the EMTs lifted him up into the ambulance waiting in the driveway, Whitney motioned for one last photograph with his dad. Father and son leaned in together. Ashley snapped the shot. Whitney gave one last look at the sky as the gurney went into the ambulance. Ron flashed him a thumbs up. Janet climbed in, and the ambulance pulled off.

Hours later, I was once again standing in a hallway, this time just outside Whitney's hospital room. I looked in through an observation window with his mother. He is handsome, as I'd seen in his photographs, with those thick, dark lashes, bushy brows, and his father's deep blue eyes. High cheekbones give him a regal bearing, but the wrinkles at the corners of his mouth show his sense of humor. Whitney, I'd been told by now, was a bit of a clown. He loved to make people laugh. Those are intelligent eyes, I thought.

After the ambulance had driven off, Janet had called me on her cell phone. She said that Whitney had asked about me. He had seen me through the ambulance window, and he wanted to know who I was. "I was able to tell him that you are writing a book about all this, and he wants you to ask him questions," she said.

I had followed the ambulance over in my Honda to El Camino Hospital as morning turned into afternoon. Whitney was now waiting in an isolation room to be wheeled to surgery. Nurses were coming and going, and checking the monitors, prepping him for the minor procedure. Whitney smiled at them and thanked them with peaceful prayer hands.

He turned his head and looked at me through the glass. And I looked back. In the harsh fluorescent lights, his skin glowed white, translucent even. A touch of unwanted gray hairs threaded his brown beard. Then the hospital room door pushed open, and his mom came out. I followed Janet into the small hospital room with the heart rate machine beeping and a notepad gripped in my left hand. And I met Whitney for the first time.

Chapter 3

The Adventure

I SMILE AND NOD, AND he smiles back. It's like watching the image I've created inside my mind of this young man, built out of bits and pieces of stories and interviews and writings and photos, come magically to life. It is easy to like him, and I do like him, right away. Suddenly, the once invisible patient who has no voice lifts his hands to speak.

Whitney points to me with his index finger, then curls it down to meet his thumb creating an O that he crosses with his left index finger creating a Q.

"Question?" I ask. He nods.

With pleading eyes he shows how desperate he is for me to understand, for others to understand, what it's like for him lying in his bedroom day after day, year after year, sometimes lacking enough energy to lift a finger to press the button by the side of his bed to call for help. Growing agitated, he reaches out and grasps invisible bars in both his fists and pounds them into place evenly spaced two-by-two around his bed. "Your life is a prison?" I ask. He nods. His head flops back, his eyes roll up in his head, and his mouth drops open. "You're like a corpse?" He nods. For many, many hours of many, many days. He pinches his white skin. He's not invisible. He's all too real.

"What do you do while you're lying in your room? Do you meditate?" I ask. He shakes his head no. He doesn't have energy to meditate. He spends most of each day using what bits of energy he has to control the pain of digesting the liquid food that gets pumped into his body traveling through his digestive system thick and slow like cement. He sleeps very little. He has frostbite on his belly from the ice packs that manage the crippling pain through most of the day and into the night. He touches the skin on his arm again and grimaces. "It hurts to be touched?" He nods, then spells out A L O N E, writing on one of the soft brown blankets that traveled with him in the ambulance from home. Our eyes glisten with tears. He plasters his face with a mask of fear, his mouth frozen in a silent scream, then spells out another word on the blanket, P A N I C, and my breath catches as I feel a fist of panic punch into my own gut.

Whitney touches the gray in his beard and shakes his head. He has lost so much time. He's missed so much. Rock bands that he might have loved. Photographs he could have taken. Stories he could have told. Elections he might have campaigned for. He's missed the deaths of loved ones, and their births. He's missed romances, and marriage, and children. He missed his sister's wedding day. His hopes lie with his dad. He nods as he spells it out on the blanket: D A D. Then he spells out: H E R O. His dad will figure it out. He draws a line across his throat and shakes his head. No suicide. It's not an option. He's living for the many others out there sick like him. If he can continue on as sick as he is, other ME/CFS patients will too. Suicide is far too common within his community, and he wants to help stop that.

"You can't talk, you can't eat, you can't listen to music," I say, sympathizing. At the mention of music, his beautiful face crumples into creases, rivers of grief. A sob escapes his mother.

But then he recovers yet again, and he continues on. This is just part of his story. It's important the rest of the story gets told too. Yes, Whitney has lived through years of hell trapped inside his broken body, but his life wasn't always like this. He was an adventurer and an artist. He had girlfriends and a favorite dog and loved ice cream. He never felt anxious or panicked before he got sick. He shakes his head. He starts to explain. In his cupped hands he holds the sphere of an invisible Earth.

"You traveled the world," I say. He nods.

I tell him I know of his travels. I know that he visited Jamaica and India, Ecuador and Guatemala, that he campaigned for Barack Obama during his presidential campaign, and that he won photography awards. That he loved hiking and nature and tending his flower gardens. I tell him that I will write about his adventures. There would be gaps in the story, but I'd work to create snapshots of his past life as best I could with words. The desperation in his face changes to some semblance of relief. And he smiles, the same crooked smile his dad sometimes has, his eyes bright.

He draws circles in the air with his finger. A bicycle? His head shakes no. A moped? No. A motorcycle? Yes, he nods vigorously and grins. "They told me you rode a motorcycle through the Himalayan mountains." His eyes crinkle into a smile. *You see? I don't want to be here,* his face says. *I want to be far away.*

I left that day utterly exhausted. As excited as I was to meet Whitney, communication had been difficult, emotional, and frustrating as hell. I was afraid it would be the only time I'd get to meet him, but I was wrong. After that day, Whitney began asking for me to come visit whenever he went to the hospital for a feeding tube replacement. He had a pantomime for me now. He would create a T with his fingers and make a book with his two hands opening and closing, indicating "Tracie—book lady." About once every three months Janet would send me a text with the date and time of the next feeding tube replacement. No matter how busy I was, I always came. I'd meet the family at home, then follow the ambulance over to the hospital.

Whitney never spoke during these meetings, nor could he read or write. At times, he could look at a photo; other times, that was too much for him. He used dramatic facial gestures and arm gestures to tell his stories. Janet called it WSL for Whitney Sign Language. Sometimes, with Janet or Ashley's help, I could guess what he was trying to communicate. Often I failed, and he'd get so upset, shaking his head and arms in anger. I constantly worried about pushing him too hard and causing him to crash. Whitney was in bad shape, but I knew he could still get worse. I also knew that he was passionate about sharing his story. He signed to me that it was worth it.

Each time I left him, I returned to my home in Santa Cruz, where I had turned my daughter's upstairs bedroom into a home office, filling it with books and files on ME/CFS. I'd get to work on my laptop trying to fill in the blanks in the stories he was trying to tell me by digging—browsing his old photographs online, looking through ME/CFS chat rooms for any signs of a post he would've shared early on in his illness. I imagined him back then, sick but still able to talk, listening to music, even watching movies on his laptop with Ashley. Now he was a grown man, stuck in a room, still cared for by his mom. How did he get so sick? Did his parents try to help him early on? Of course they did. His dad wasn't a medical doctor, but still he worked at one of the most prestigious medical schools in the world. Why couldn't he find answers? I was trying to make sense out of all of it.

One day I read an essay Whitney had posted on his photography website in which he was obviously trying to do the same thing. Make sense of it all. He was twenty-nine years old, getting sicker and sicker. He could still type on his computer, apparently, but only from bed. Obviously, he was scared to death about what his future held:

"Really sick," he wrote. "I can't talk. Can't type/text enough to communicate. Haven't had a conversation with someone in six months. I have been struggling with health problems for the past eight years since I was 21. Every time I traveled my health seemed to plummet. But I have always been inspired and dedicated and never thought I'd wind up where I am now. So I kept going, kept pushing myself to do everything I wanted to do. My trip to India was the last straw it seems."

Whitney's trip to India was the pinnacle of his traveling adventures: this he had indicated to me many times through his hand gestures. He was twenty-three years old at the time, on an extended break from art school, planning to stay for months, or at least until his money ran out, but with no real schedule in mind. He flew from San Francisco into the Himalayan mountains, landing in the region of Ladakh, known as the "gemstone" of the Himalayas, popular with tourists for its raw beauty and the rugged mountains, its many monasteries and Buddhist monks. Whitney was drawn to Buddhism and planned to study it while he was there. Parts of the region soar to altitudes of eighteen thousand feet, peaking at unbreathable altitudes of twenty-three thousand feet—a height that rivals neighboring Mt. Everest's twenty-nine thousand feet.

As soon as he landed, he got sick with a strange lightheadedness that made him really dizzy, with stomach pains and fatigue. He hoped it was just altitude sickness, which commonly occurs at altitudes as low as eight thousand feet. Still, he was worried. He'd been getting sick with similar symptoms for the past two years, ever since a trip to Jamaica. But never this bad.

As I read the emails Whitney sent to his mom at the time, I had been a bit shocked to learn how sick he had actually been, even before he got to India. And also surprised that he almost terminated his trip soon after landing there. He wrote to his mom that he had scheduled a return flight before the end of the month, and he had two doctor's appointments planned for right after he got home.

"I just want to put an end to this once and for all," he wrote. "It has dragged on and tarnished enough of my life already.... I want to put all my energies into seeing doctors. I'm sick of only being able to give 50 percent."

And then, like a typical adventurous young guy, he just changed his mind. He didn't want to go home, and anyway, when he traveled to lower altitudes, he began to feel better. He never got on that flight and instead stayed in India a total of nine months. He emailed home that he felt fine, that he was having the time of his life backpacking, riding motorcycles, taking buses or trains across the country. He wrote about plans to ride a jeep over one of the highest motorable passes in the world. "Like 5,000 meters or something crazy..." And camping by a "crystal clear waterfall," like when he was a kid backpacking with his family. He helped build a Buddhist nunnery and took a bus and a train to Nepal near Kathmandu, where he lived at a monastery for five weeks for an intensive Buddhist philosophy retreat.

"The poverty is sad," he wrote. "But the spiritual community here is really beautiful." His emails were spotty, and despite his mother's pleadings he didn't return home for Christmas. He wrote about being thrilled to experience an "anti-commercial Christmas." And then suddenly his emails stopped. Three months later, he was in a hospital in Calcutta, deathly ill.

Searching for answers to fill in that three-month gap, I returned to Whitney's virtual footprints online, but I found nothing. Eventually, I discovered the rest of the story in a document filed along with all of Whitney's medical records stored in a chest in his parents' dining room. Whitney had written it several years after his trip to India, as a kind of medical diary for his doctors and his dad, who were still trying to figure out what was wrong with him at the time.

Five months into his trip to India, Whitney got really sick again. One night, about the time his emails home stopped, he had nausea and mild diarrhea. For the next two weeks, he was incredibly ill, with exhaustion, upset stomach, and fever. He stayed sick for two weeks, got better for two weeks, then got sick again. He must have seen doctors there. But he refused to go home and kept pushing himself on, determined to live out his grand adventure.

"For three months, I took medicine for parasites and worms and maybe an antibiotic," he wrote, admitting he was so sick at that time the details were hazy. "After three months of this I had lost 40 pounds, weighing 125 pounds. I developed a fever and pneumonia and was hospitalized in India." Then he came home. I asked Whitney once what the hospital was like for him in Calcutta, but he just grimaced and looked away.

When Whitney arrived home, he was a skeleton of himself at 115 pounds. Ron and Janet were shocked. He was tormented by severe dizziness, along with debilitating headaches, swollen glands, fevers, and stomach pains. His energy level had dropped to zero. Ron suspected that he had picked up parasites on his travels. Ron told me that sometimes parasites are really good at hiding, so even though Whitney tested negative, doctors treated him anyway. The vertigo, or whatever it was, slowly dissipated, but the punishing fatigue remained. Routine lab tests all came back negative. One doctor told him there was nothing wrong with him.

"At first, I thought it was an intestinal parasite," Ron explained, but the medications didn't help. "Eventually, I talked to one of his doctors in India, who said his lab tests were all negative. We tested for Lyme disease multiple times. That came back negative." Whitney, growing more frustrated with Western medicine, then suddenly disappeared again. A friend in India had told him about this natural healer, a shaman in Guatemala who could cure him. It was a short trip.

"The trip was a disaster," Ashley told me one time when we met for coffee in downtown Palo Alto, at one of those Silicon Valley wide-open rooms, filled with computer scientists tapping away on laptops. "He grew sicker and sicker, until he was stuck in some fleabag hotel in

Guatemala City. He was stuck there for twelve days until he got better enough to come home. Whitney's still pissed at himself for going. He got so much sicker."

Back in San Francisco, a new round of doctor's appointments began in earnest. Ron insisted he get an endoscopy and a colonoscopy. Results were negative. Whitney's health would go up and down, but he refused to move home with his parents. At some point, he got a temporary job working as a paid organizer in Carson City, Nevada, for then Senator Barack Obama's presidential campaign. Obama was one of his heroes. His health still wasn't good, but he continued to push himself.

Finally, Whitney had to accept that he could no longer live the life he longed for, one of travel and adventure. He left school and relocated to Berkeley to set up a part-time wedding photography business, leaving time to figure out how to get better. His business did well at first. He set up a website and began to draw clients. He liked the work, and his clients liked him. They appreciated his attention to detail, his storytelling skills, his open personality. He won more awards for his work and was able to maintain some semblance of the independence he longed for. But it was getting harder. The twelve-hour work days photographing weddings sent him crashing sometimes for a week, meaning he could barely get out of bed.

Any leftover energy went into navigating the medical scene. He visited one physician after the other, racking up scores of appointments and blood draws and medical costs.

All tests came back negative. Physicians sent him to psychiatrists. The psychiatrists sent him back to the medical doctors. He tried all kinds of natural remedies, all kinds of changes to his diet, medications, exercise, lack of exercise. Nothing helped, and no one had any answers. His frustration levels climbed. Cardiovascular exercise, in particular, sent him crashing. And even when he could sleep, he felt no better.

"Initially, I still thought the doctors should be able to figure this out," Ron told me. "I supported him and took him to appointments when he could no longer drive. A few told him he was imagining he was sick. That he wasn't eating because he didn't want to. One said he could cure him with vitamin C. Another told him his jaw was pressing on his nerves, causing all of his problems. I told him to stop listening to these fringe doctors. But he was desperate for answers."

Daily living became a challenge; cooking a single meal was a struggle. His sister would stop by to help him clean when she could. Friends came over to wash his dishes. But as time passed, friends grew frustrated with him and slowly began to fade away. He was very thin but otherwise looked fine, and they just didn't get how sick he was. In reality, Whitney barely had enough energy to get through each day.

And then he didn't.

On November 8, 2010, Whitney posted on his wedding photography Facebook page: "So you may have noticed that I haven't been posting weddings lately. I've gotten really sick with something I picked up while traveling in India, and I am going on sick leave until I get better. So no weddings this coming summer unfortunately." There would be no more weddings for Whitney, not even his sister's. For three years, he'd been searching intensely for answers, and still he had no idea what was wrong with him.

Chapter 4

A Mysterious Diagnosis

On a day in August 2010, Whitney made yet another doctor's appointment, this one with someone new in an office across the San Francisco Bay from Berkeley in Marin County. I liked to imagine Whitney driving across the Richmond–San Rafael Bridge that day with the radio blaring. He was quite ill but still able to drive himself places at the time, determined to keep his independence, determined to find answers. I had driven over the same bridge many times during my college days to visit a boyfriend in Marin County and remember its stunning views of the San Francisco skyline to the left and the notorious San Quentin prison sitting on the rocky shoreline up ahead. I hoped the view brought Whitney a bit of peace.

The appointment was in Santa Rosa with Dr. Eric Gordon, a family practice physician and one of those rare doctors who specializes in the care of patients with chronic diseases. In particular, he focuses on patients who are sick but, for whatever reason, whose routine lab tests all come back negative, showing nothing wrong. This usually meant patients with autoimmune disorders or diseases like fibromyalgia and what Gordon was referring to at the time as CFIDS (chronic fatigue immune dysfunction syndrome) aka CFS, yet another name for the disease. That day the doctor sat and listened to Whitney describe his years of illness, the unrelenting fatigue, the brain fog, and stomach pain he'd been living through. How his fatigue worsened with any sort of exercise. The doctor nodded his head in understanding as Whitney talked. He'd heard similar stories many times from patients. At some point, he scribbled CFIDS in the margins of Whitney's medical chart and then told Whitney he knew what was wrong with him.

The first time I met Dr. Gordon was at the first of Ron's international ME/CFS conferences held on the Stanford campus years later. Ashley was moderating. Janet sat up front with Ron.

The conference attracted scores of CFS patients, their caregivers, and family members looking for hope. There were patients in wheelchairs and others in special reclining chairs designed to prevent fainting and nausea. Heads rested on shoulders of loved ones. A tiny service puppy on a plaid blanket curled up on a middle-aged woman's lap. During breaks, patients stretched out on blankets outside on the campus lawn. Janet brought lounge chairs from home and set them up outside. Many, many more patients, too sick to attend, listened to the conference streaming live online.

During a break in the scientific presentations, I spotted Dr. Gordon chatting in hushed tones with Dr. Dan Peterson, whom I recognized as one of the two original Incline Village doctors at the center of the outbreak on the shores of Lake Tahoe three decades before. I went back and joined them, introduced myself, then, with their permission, listened in.

"When I apply for grants I never call it CFS," Peterson said. Gordon nodded. For three decades, prejudice and discrimination toward the illness have continually roadblocked their work, they told me. They'd been forced to learn work-arounds.

"Do you think attitudes are finally changing?," I asked. They both chuckled softly.

"We will have to wait and see," Peterson said.

During lunch, I sat next to Dr. Gordon and asked him if he remembered that first appointment he had with Whitney.

"I remember it well," he said. "He was easy to diagnose." Dr. Gordon was a kind man, I could tell right off, one of those doctors who really care about their patients and put the extra time in to listen. He wore glasses and spoke gently, often using his hands to express himself. He'd spent a career listening to patients with chronic illnesses and was a good listener.

"People with post-exertional malaise, who are otherwise fairly healthy, usually have this disease. Other diseases, if you rest, you feel better," he explained. Different traumas can trigger the disease, he said, like infectious diseases or even car accidents. "I've seen a lot of patients like Whitney get sick the first time in India or somewhere in Southeast Asia. It usually sounds like a parasite triggers it, and I think it often is." Then he repeated himself.

"He was easy to diagnose," he said. "But I couldn't help him."

Not long after his diagnosis Whitney was finally forced to give up his independence. He just couldn't take care of himself anymore. He moved out of his rental in Berkeley and settled into the back bedroom in his childhood home, the one with the view of the pool and a shady oak tree out back.

Over the years to come, squirrels would gnaw on the wooden porch that leads off the bedroom, the annoying noise painful to him. Cooking smells would drift in under the door from the kitchen, causing him to crash. On a crisp fall day, two or three wasps would sneak inside from a nest burrowed into his closet from a crook in the outside wall, tiptoeing slowly across his sunken, bare chest. Whitney would lie still and watch them, amused. From his bed, he would gaze out at the African daisies and delphiniums he once planted in the garden during happier days, as they bloomed, over and over and over again.

The internet was now a lifeline, somewhere he could research his strange, new illness and connect with others like him. He searched for new treatments to try and more research to investigate. He asked for advice from patients with ME/CFS in chat rooms and found some solace for the growing loneliness that would soon come to envelop him.

While having a diagnosis had brought him renewed hope, learning just how little was known about the disease, and the lack of official treatments, began to dampen it. He was learning how his illness had been ignored; how patients suffered, not only incredible physical pain, but the mental anguish of not being believed, of being mocked and stigmatized. And how many were simply abandoned.

When he first moved back home, Whitney's fatigue was worsening daily, and excruciating pains in his legs began to hobble him. In one chat room, he posted: "I saw my GP the other day and he said he wouldn't help me get a wheelchair covered by insurance because he wouldn't feel comfortable with it because he thinks I should walk.... I was like, 'Hey, I think I should too, but that doesn't change anything.' %**##$."

Everything Whitney learned, he shared with his dad, the smartest person he knew. His dad always had the right answer to difficult questions. And Whitney's faith in Ron was unshakable. When Ron came home in the afternoons to help care for Whitney, the two spent hours talking trying to figure out answers. Whitney shared the stories of others he'd come in contact with. Often their stories were tragic and, in fact, scared the crap out of him.

As I read through Whitney's comments in these chat rooms, I would come to hear one thing repeated over and over again: to die of this illness is atypical, but to hover in an in-between state experiencing a "living death" for years or decades is typical.

Curious to hear the story of how Ron first decided to take on researching the disease himself, I asked him about it during one of our meetings in his office.

"What made you take the leap and start your own investigations?" I asked. He grew thoughtful.

"I was stunned by the lack of medical research on the disease," he told me. "Then I checked into how much NIH money was available to fund research." As an experienced researcher—one who had applied for and been granted many multimillion-dollar research grants from the National Institutes of Health—he expected there to be a certain amount of money for the numbers of patients. In 2011, while diseases like multiple sclerosis, with similar-sized patient populations, received $100 million per year, CFS was getting only $6 million. The amount was so small that he immediately understood why no one studied it. Scientists couldn't afford to.

Meanwhile, Ron began to work with Whitney to find new treatments for his symptoms. He read through all Whitney's medical files from his two new doctors, Eric Gordon and Andy Kogelnik, both experts in CFS.

Whitney had tested positive for several virus antibodies, including the Epstein-Barr virus and the Human Herpesvirus 6, HHV-6. His immune system showed abnormally low activity levels of those so-called natural killer cells that fight off viruses and cancer cells. His immune system wasn't working right. Exactly why, though, no one knew. He was prescribed various antiviral medications, including Valcyte and acyclovir. They were expensive, but Ron agreed that his son should try them.

They tried the experimental treatment rituximab, a drug therapy for certain cancers, which was shown by happenstance to help a patient with ME/CFS. The drug cost $7,550 per infusion, much of it not covered by insurance. Treatment involved getting infusions at Dr. Kogelnik's office over a sixteen-month period. It didn't help. His doctors prescribed different antibiotics, anti-inflammatories, supplements, and antidepressants. None of them helped. And Ron was beginning to lose faith that any doctor could help his son, even experts in ME/CFS who were doing their best with the limited options available.

At the same time, they tried all kinds of alternative medicines—acupuncture, Chinese herbal medicines, Tibetan medicine, and a Native American healing ceremony. They also consulted a mold expert after hearing that some ME/CFS patients were particularly vulnerable to mold. Some things helped a little. Most things didn't.

Ron would come home in the afternoons and help Whitney put on compression socks to ease the pain in his legs. And continue to add the endless numbers of supplements and antibiotics and other drugs to his IV line. Whitney was diagnosed with sleep apnea and tried using a CPAP machine to help him breathe. After months of trying, he gave up. He just didn't have enough energy to deal with it. Ron began to panic. He watched his son wasting away and couldn't do anything to stop it.

Whitney still managed to find bits of joy listening to his music: he filled the empty hours by creating playlists on his iPod from bed. He liked to curate the lists to tell a story and named them things like *Long Road Mix* and *Bummer Mix*. He could eat some foods, although that was getting more difficult each day. Ice cream had been replaced by yogurt and constant stomach pain. But Ron was busy hatching new plans. He couldn't stand by and watch his son's endless suffering.

And so he got started on his own. I imagine him one night, shuffling out to his toolshed out back after setting up Whitney's IV, filling his water containers, and changing his socks. Ron had set up a sort of makeshift science lab on his tool bench. It was cluttered out there, so he cleared out a space on the workbench, pushing aside his old woodworking tools. He showed me the centrifuge—a tool used for blood separation and analysis. It's still there, small and round, rather old, but functional. It reminded me of one I'd used in a high school chemistry class.

"Is this where you built Whitney that beautiful oaken cradle that you guys keep inside the house?" I asked him.

"No, we hadn't moved to this house yet, but these are the same tools," Ron said, looking at one of the hammers nostalgically. He hadn't been out here for awhile and was sort of embarrassed by the clutter. Plus he was always hesitant to talk about himself, so he laughed nervously when he added, "My dad was a carpenter. I'm good with my hands like he was."

That's how his scientific investigation first began. Once a week, after chatting with Whitney in his bedroom, he'd take a vial of his son's blood, then carry it with him to the work shed out back, curious to see if he could find any molecular clues to the mystery. He'd watch it spin around fast in the centrifuge, separating into its different parts, and then he or Janet would get into the car and make the twenty-minute drive over to his lab for processing.

This is the way Ron has worked throughout his life. He would find the right tools to tinker with, in settings where he felt free to disappear into the imaginary three-dimensional worlds of scientific exploration. A place that feels safe to him, that feels like home. This is how he made his first big scientific discovery.

In 1970, his final year as a grad student at Caltech, Ron was in a cool, dark basement where he spent months tinkering around with the electron microscopes. They were huge at the time, the size of refrigerators, and rare—hard to get your hands on.

"I hated the hot summers in Pasadena," Ron told me once. "So when it came time to choose a project, I picked a topic that entailed using the electron microscopes that needed to be kept in the basement for the cool temperatures." He spent hours on end, days at a time, down in that basement, eventually cracking one of the most tricky genetic puzzles of the times, and pioneered a method for the physical mapping of the location of genes along DNA, one of the first mapping methods for DNA.

As Ron began to study his son's blood cells, he also began to worry about how he would fund any future research. He set to work making plans for experiments and more advanced testing, getting his lab involved. He knew it would take a lot of money. He made plans to run every kind of available test in his high-tech Stanford lab on Whitney's blood, searching for clues of what had gone so badly wrong in the cells' molecular pathways that could lead to treatments or even cures. The list was long and complicated. Testing would include things like cytokine analysis, genome sequencing, microbiome sequencing, metabolomics, magnetic levitation profiling, PCR assays for any viruses, antibody assays for mycotoxins, and much more.

Over the years, Ron's lab had developed a wealth of biotech inventions and advanced diagnostic testing tools. In 1989, Ron cofounded the Stanford Genome Technology Center with a large government grant to help build tools for the $3.8 billion Human Genome Project, the same project called by President Bill Clinton at its completion "the most wondrous map ever produced by humankind." Ron became director of the lab in 1992 and has remained there since. The lab made a name for itself as a think tank for the creation of diagnostic tools to help battle human illness and pinpoint disease. It also became the launching pad for biotech scientists who would go on to develop successful new startups to advance medical care.

But now he was thinking about changing the course of his research. Exactly how to launch this new project kept him up nights. He needed a plan.

One afternoon, during a visit home from Hampshire College, Ashley was pushing Whitney in a wheelchair around their quiet Palo Alto neighborhood. It was a sunny spring day, the air perfumed by jasmine and wisteria. She couldn't help but see all the beauty that was still alive in her sick brother. She planned to move home from Massachusetts after graduating and take care of him as much as she could.

The two had grown up best friends, and they would be forever. Her heart would always be with her brother. She was angry and scared about his condition, but still he inspired her. Living her life without Whitney in it was unimaginable. He had always had her back. Ashley was five years Whitney's junior. Forever the little sister, she thought her older brother was the coolest ever. Still did. He was a gentle soul who never could hurt anyone or anything, not even a spider, she told me. He stuck up for little kids, saved abandoned animals, and befriended the misfits in high school. He was handsome, and the girls buzzed around him. When they were little, he used to tickle Ashley mercilessly and wrestle her down to the ground, and they'd both laugh hard. He was funny and fun and always in constant motion. Ashley would laugh, a bit wistfully, thinking back to their many hikes together. She missed so much about her brother. But she would hold on to all the parts of him that she had left. Ron had visited her at college when he was in Boston at a conference. He talked to her about his plans, and she knew they needed funding. Maybe she could help with that. She didn't know much about raising that sort of money. But she knew that her dad was brilliant and that somehow they'd figure it all out.

That afternoon, as she was pushing Whitney through the neighborhood, she started talking.

"I think your story could touch people," she said to Whitney. He didn't respond, but she could tell that he was listening. Her brother had always listened to her. He never made her feel like the annoying little sister. "You know, like celebrities do?" She could see him as like Michael J. Fox, the way he had raised so much awareness about Parkinson's disease. It wasn't only that she loved Whitney so much. Her brother could walk into a room and charm everyone. She was always proud of him. Just by being himself he seemed to be a magnet to others. She was more shy, but he'd light up a room. She knew her brother could never be invisible. They could raise millions of dollars for research. They could find a cure for his illness. She believed it with all of her heart.

She also knew Whitney was an intensely private person, and taking up the call to share all the details of his private life would give him pause. But, already it was too late for privacy. Whitney was a full-fledged ME/CFS activist by now. His mission was to fight against this disease and all those who denied it. Many times he'd shake his fist hard, to show that this was his fight. From the time he got diagnosed, she knew that he wanted to help the millions of others who were sick, not just himself.

"You know?" Ashley said again as they continued down the block. And he nodded in agreement.

The kitchen transformed into a pharmacy. The table overflowed with an arsenal of medications. A sweet, antiseptic smell crept into the air and stayed there. Whitney's handwritten notes still remain stuck to the closed door of his bedroom from those years: "I don't know what to say. I just feel pretty hopeless. It's so hard not being able to take care of my stuff."

The list of Whitney's symptoms seemed to grow longer and longer the more Ron looked at them, which isn't unusual for ME/CFS patients. Some say the list of possible ME/CFS symptoms can reach up to sixty. Whitney listed his once: dry eyes, inflamed gums, hair loss, digestive sensitivity, freezing feet and legs, leg pain ("It seems specific to walking; crawling doesn't seem to trigger it."), shortness of breath, lightheadedness, swollen glands, sore throats, poor sleep, and ever-increasing extreme fatigue.

"I have an energy 'envelope' and can think fairly clearly and do very minimal physical movement (like moving around inside my room) for a few hours until my energy runs out and I have to rest because my body ceases to be able to function," Whitney wrote in an online posting once. "If I push it past this point it can take days to recover, so I don't do that anymore. But I cannot even talk to people or work on a computer for more than two or three hours at a time."

Then, finally, some good news. Ron was awarded one of the most prestigious prizes in genetics, the Gruber Prize, an international award founded at Yale University. Recipients are selected by a panel from nominations sent in from around the world, and he was being honored for his "pioneering work in genetic engineering." The award pleased him, but he'd received many before, and awards really weren't important to him. The reason this award was important this time was because it came with a half-million-dollar cash prize. And he needed money. Here was the windfall he needed to get his ME/CFS investigation off the ground.

By 2013, Ashley had officially set up the website for the new Chronic Fatigue Syndrome Research Center at Stanford. Ron now began to turn his full focus to ME/CFS research. It was an unexpected turn for a senior scientist noted for his years of genetics research. He immediately started to recruit some of the young scientists and colleagues in his lab. And he came up with his plan.

So far, there was very little data at a basic molecular level that would allow him to generate any hypotheses about the underlying process for this disease, he told me.

"The scientific process begins with observations," he said. "So I needed lots of them. I planned to cast out a large fishing net, then see what fish I caught." That meant gathering together billions of molecular observations and looking for clues. From those clues he could generate hypotheses to test.

"I wanted to find something that would help the patients," Ron said. "Treatments, diagnostic tests, and hopefully a cure for all those people suffering, including my son."

Chapter 5

Rocket Boy

THE LOBBY OF THE Stanford Genome Technology Center is staffed by a security guard, and the walls are covered with neat rows of framed patent certificates, dozens with Ron's name on them. To get to his office, you turn left past the guard and left again into a small room with an assistant's nook up front and a large desk in the room in the back. Behind this desk, Ron keeps a bookshelf filled with genetics and biochemistry texts, along with several large green volumes of the hundreds of his original research articles published in the most prestigious of the world's science journals—*Science*, *Nature*, the *Proceedings of the National Academy of Sciences*, and more.

Up on the wall across from his desk, where he can be sure to see it every day, hangs a framed photo of a healthy, tall Whitney, looking mischievously down at his dad standing next to him. Whitney looks to be in his early twenties, and he's got Ron's wire-rimmed glasses perched low on his nose and is stroking his whiskered chin in a professorial manner. (The same motion he uses as a pantomime for his dad now that he can't speak.) He's parodying his dad, and Ron looks out at the camera with a wide, wide grin.

"Ah, he's mimicking his dad," I said the first time I saw the photo during one of my many visits to his office.

Ron nodded. "Oh yeah, he was always joking around."

As I grew more immersed in my research about Whitney's story, I'd walk across the street from my cubicle in the communications office, usually on lunch break, to Ron's office to say hi and chat about his latest research. Getting him to return a phone call, or even find your email among his thousands, was pretty much impossible. But I figured out all I had to do was make an appointment with his assistant Katrina several weeks in advance. Then he'd have some time to talk, and he always greeted me with a smile. Back in 2016, when I first visited him, I discovered his office was also a place where he felt comfortable opening up and talking about himself a bit more. He was quiet at home. Mostly, though, he just wanted to talk about science and how it could help save his son.

On the first such meeting, I found Ron sitting behind his desk, preparing to drive home for an afternoon session caring for Whitney. He was wearing his jeans and sneakers, with the North Face logo sanded off. (Labels on clothes bothered Whitney.) But he said that he could make some time for me before he had to leave.

"We're sequencing Whitney's genome from his blood samples, looking at expression of the white blood cells," he told me, launching right away into the science. White blood cells are immune cells, the cells that fight off infection. That sounded good, I thought. He was studying what these cells were producing—or not producing—in Whitney's body, like antibodies, for example, to fight off any invading infectious cells. Often, when Ron talked about science, it would fly so far over my head, I'd squeeze my eyes tight to keep away the tears of frustration, as if I were a kid stuck on a tough math problem. I'd been working for more than a decade as a science writer at Stanford, but I wasn't trained like my coworkers with science PhDs. For me, it was often a challenge understanding Ron.

When I asked him how things were going, Ron admitted he was distracted. He was trying to keep thoughts of Whitney and suicide at bay that day. He tried to focus on the science, but those thoughts kept intruding.

"I have a lot of respect for Whitney," Ron went on. "He's had friends with this disease who have killed themselves." Ron told me about a study that showed nearly a sevenfold increase in suicide among those with ME/CFS compared to the general population. Then he looked up at me, with hope in his eyes. "But Whitney told his mother that he never would because of his Buddhist beliefs. They won't allow him to kill any living creature. He took a vow."

Ron smiled then, remembering back to Whitney as a kid when he used to call his dad into his bedroom at night to gently pick up a stray spider and take it safely outside.

"I like to think that he travels in his mind to all the adventures he lived before he got sick," Ron continued. "That he can travel far away in his mind, maybe create new worlds. That's something I learned how to do as a kid. I was always sick as a kid."

Ron was a quiet, slight child with intense blue eyes. Born on July 17, 1941, he was the third child of Gerzella and

Lester Davis, a hard man with a sixth-grade education who was proud to say he supported his family by the sweat of his brow as a tradesman and that higher education was a waste of time. Before his first birthday, Ron contracted a serious case of rheumatic fever along with his older sister Patty. But while Patty recovered relatively quickly, it would be years before Ron did.

By the time he was six, Ron had suffered through dozens of bouts of strep throat with high fever, the persistent effects of his rheumatic fever reattacking with a vengeance every few weeks and forcing his parents to pull him out of the first grade. Nothing doctors could do ever seemed to help. They lived in Charleston, Illinois, a farming town with a population of eight thousand, so there weren't many options for getting care. Ron was forced to suffer the constant strep throat, high fevers, and earaches with little hope of a remedy. In his family's small two-bedroom house he would lie on the couch, too sick to think or talk, utterly miserable for weeks at a time, sometimes months, year after year.

Ron would stay home with his mother. He said that she was a kind woman, with a seventh-grade education, who didn't like to wear shoes. She always kept busy and often talked to herself as she did her chores, sometimes ironing shirts she took in from the community for ten cents apiece. On one particular day, a day forever etched in Ron's memory, his mom was busy like usual, most likely washing the dishes or keeping the oil furnace in the living room burning, when the front door clicked open, and the family doctor, Dr. Icknan, came in for a routine house call. He was a gray-haired man with a friendly face, and he walked over to the couch carrying a black doctor's bag.

"He leaned over me and said, 'I've got something new for you this time,'" Ron told me.

A scientist named Alexander Fleming had already invented penicillin by then. It was the first of many future antibiotics that would change the face of medicine and save untold lives. It could have cured Ron, but prior to 1949 in rural Illinois many doctors like Dr. Icknan couldn't get their hands on any of it—until that day, which is why the country doctor was smiling from ear to ear.

Ron was born just six months before the Japanese bombed Pearl Harbor, launching the US into World War II. The new infection-fighting drug had been immediately put into mass production and sent off to soldiers battling infections from gunshot and shrapnel wounds, but leaving limited supplies back home in the United States. Dr. Icknan would have to wait for a few years after the war ended before it made its way to the small town of Charleston and safely into his black bag.

Ron fondly remembers Dr. Icknan reaching into his bag and pulling out a small glass bottle. Twenty minutes after the medicine was injected into his arm, his fever broke. He was up walking around. He could swallow without pain. It was a miracle.

"Boy, that made a big impression on me," Ron said. "I remember thinking, *Oh my gosh, medicine can do this? Then this is what I have to do. I have to make things that can make sick people better.*"

Still, for Ron the penicillin came too late to completely cure his illness. It just made him feel better each time he got sick. It was a temporary solution. By the time he was twelve, he had come down with more than two hundred bouts of strep throat. The rheumatic fever gradually receded over the years with rest and penicillin, but by then it had left Ron with a permanently damaged heart.

After six months of bedrest, Ron returned to school, but he was still the sick kid, thin and frail, and he became a constant target for bullies both at school and at home.

Gene, his older brother, was the opposite. He was a strong athlete who, in spite of being short, played basketball. He drove fast cars and teased Ron mercilessly, constantly punching him in the arm, until Ron would climb onto the roof to hide out under the branches of a leafy tree.

His father's beliefs didn't help. Ron, like his brother, was expected to help out with his dad's hardwood floor business, cutting and laying wood planks, then sanding them until they shined. Once, when he was fifteen, his dad dropped him off at a high school gym, handing him a crowbar, an army cot, and some cash.

"Tear out the water-damaged floor and call when you're done," his dad said.

"I finally called him after several weeks," Ron told me. "I was hot and sweaty, covered in splinters and bug bites and my back ached. When he picked me up, all he said was, 'Good job, but it took way too long.'"

His father constantly reprimanded him.

"You'll never amount to a hill of beans," he told him repeatedly.

Ron's father, tall and fair like him, believed the only honest way to get ahead in the world was through skilled labor as a tradesman. Lester Davis left school in sixth grade. He was born poor in an Illinois log cabin. He left home early, finding labor as a migrant farmworker hopping empty boxcars to follow the crops from Texas to the Midwest and back again. When he met and married Gerzella, they settled down in Maroa, Illinois, population 1,000.

When the family moved to Charleston, Ron was two years old. His father was looking for clientele for his new business as a carpenter hand-crafting hardwood floors. He was a man of strong opinions. He hated President Harry Truman, called TV anchor Walter Cronkite a "commie," and once came after "union folk," who were demanding he join them, with a four-by-four. He had no patience for the professors who hired him to build their floors. They had book sense but no common sense, he'd say. That was no kind of profession.

Decades later, when Ron grew up and gained notoriety as one of the world's most brilliant scientific minds, his family in Illinois never quite understood what it was he did. His mother once told Janet, "It sure is a good thing Ronnie doesn't have to work for a living."

If anyone had told his family back then that Ron was a genius, they would have laughed out loud. Ron struggled in school early on. He'd missed so much school when he was out sick that he constantly had to play catch-up, and he struggled hard to read. No one knew for many years he had learning disabilities—dyslexia and an auditory processing disorder—that make it difficult for him to process words and letters. Kids teased him. Teachers chastised him. "I always loved learning, but I didn't like the indifference that most of my teachers had for me," Ron said later. "They thought I was stupid, and I wasn't worth their time."

He believed it when so many told him he was dumb, and he often felt like a misfit.

On Sundays, he'd attend the First Christian Church with his mom and sister and would have to keep an eye on the school bullies in the congregation. Friday nights the family went to basketball games, and Saturday afternoons were matinee showings of Westerns at the Will Rogers movie theater downtown. His father loved hunting, guns, and cars, while Ron liked airplanes and walking in the woods. He hated hunting. His sister Patty sometimes tagged along when their dad and older brother went hunting, but not Ron. He couldn't stomach it. (He's the rare medical scientist who has never dissected an animal, unless yeast counts.)

Ron preferred wandering in the woods alone, studying the flora and fauna, and digging for Native American artifacts. He wasn't into hanging out at the drugstore drinking Cokes after school—that was for the popular crowd. His mom insisted on piano and accordion lessons, but what he really wanted to do was play baseball. He practiced his pitching skills endlessly at home, throwing a homemade ball of paper wrapped up in tape at a small target. Once, finally given a chance to pitch on a real baseball field, he struck out all the "hot shots" with his deadly aim. But instead of rotating him into the game as pitcher, when he even got the chance to play at all, the boys stuck him in the outfield.

When baseball didn't work out, Ron, who like his dad was good with his hands, turned to making model rockets. He spent hours in the basement mixing up chemicals for fuels that would give him the biggest bang for his buck. When the occasional explosion boomed out, sending smoke streaming from the basement, it only further confirmed for the family that scrawny Ronnie was more than a little "odd."

"They thought I was a bit nutty," Ron told me.

Ron's passion for building rockets had nothing to do with building the body of the rocket itself. That was just a thick paper tube with some fins attached. The fun part was mixing up better and faster rocket fuels. At first, Ron did what most of the other teenage rocketeers were doing back then. He'd get a box of bullets, cut them open, and use the gunpowder inside for propulsion. That worked pretty well for a while, but it wasn't very powerful, and the neighbors started complaining about the smoke coming from the basement window. The occasional explosion or two didn't bother his mother so much. She'd shrug her shoulders and say, "Oh, that's just Ronnie in the basement with his rockets."

He wanted to figure out a way to avoid explosions if possible, so he did more research and started reading in *Popular Mechanics* magazine about other chemicals that worked better and safer and still pushed his rockets faster. It was easy back then to buy chemicals like potassium nitrate for a few dimes at the local pharmacy. But then he got bored with that and came up with other, bigger ideas for even more powerful rockets. When he went to the nearby university library to check out chemistry books, but was told, even after he begged and pleaded, that he couldn't because he was only fourteen, he came up with another plan.

One summer night, Ron grabbed a flashlight and tiptoed out the front door of his house. In the dark, he walked through the quiet streets of Charleston to Eastern Illinois University, his future alma mater. Picking the lock to the university's Booth Library—Ron always had a knack with mechanical devices—he proceeded to spend hours reading by flashlight. Hiding between the book stacks inside the awe-inspiring Gothic walls of the immense library, he learned how to make the strongest-powered rocket fuels of all by reading the serious science journals, like the *American Chemical Society* and others—many of the same journals where he would publish hundreds of his own studies someday. He kept coming back for years.

"That's where you could learn about the real thing, reading about new research from real scientists," Ron said. "I started reading a little more about chemistry and realized I shouldn't be using potassium nitrate for fuel. I should be using things like ammonium perchlorate, but that was too explosive, so they wouldn't order me that stuff at the local pharmacy."

So the next day he headed over to the print shop downtown to design his own invoices with the heading "Ronnie Davis Company" and sent orders off to chemical companies in Chicago to get the real stuff. When one day a big rig pulled up to Davis's small home with a delivery, Ron hightailed it out of the house fast.

"I ran outside and told the delivery man I would make sure my dad got his delivery," Ron told me, laughing. Then the old professor held up all of his wrinkled fingers and wiggled them in the air. "I loved making rocket fuel and, look, I was good at it. I've still got all ten of them."

With his new access to chemicals, Ron proceeded to experiment with a large number of rocket fuels. The culmination of this was his creation of a very powerful fuel that he was never able to improve upon. Years later, as a researcher attending a meeting of the American Chemical Society, he started chatting with a chemist telling him about how he made this particularly powerful fuel years ago when he was just a kid. "That's the fuel that's used in the Poseidon ballistic missile, and it's classified!," the chemist said, dumbfounded. He was referring to a US submarine-launched ballistic missile developed in 1971. "When did you do that?"

"In 1955, when I was 14," Ron replied. The chemist just shook his head in disbelief.

Over the many nights he spent reading journals in the dark, Ron had discovered another chemical compound powered by nitrogen like his rocket fuels that caught his interest, and slowly a new fascination began to take hold.

"That's when I stopped making rockets and started reading about DNA and about Watson and Crick, who discovered the structure of DNA," he said. Nitrogen is one of the building blocks of the four bases that are part of the DNA molecule.

The year was 1955. During the decade following the end of World War II, what Ron didn't know was that a revolution was occurring in the world of science. Almost a century had passed since the discovery of units of heredity called genes had been made by the monk Gregor Mendel in 1865—in experiments crossbreeding pea plants. Until the 1940s, little more had been known about how these genes worked. In the years after the war, the federal government took note of the power of science to keep the nation strong. And federal health agencies like the National Institutes of Health emerged with federal dollars to spread to researchers at home and abroad.

The NIH actually traces its roots back to 1887, when it was just one institute housed in a single room before moving to its current campus in Bethesda, Maryland, a former estate. In 1940, President Franklin Delano Roosevelt, speaking from the steps of the most stately of the campus buildings, dedicated a new cancer institute and ushered in an era that would expand the institute into today's twenty-seven institutes of health.

On the verge of the US entering World War II, Roosevelt gave a prescient speech to the country from those steps, spotlighting the importance of public health to the security of the nation.

"We cannot be a strong nation unless we are a healthy nation, and so we must recruit not only men and materials, but also knowledge and science in the service of national strength," he said into a cluster of microphones. He then offered prophetic words of warning on the future dangers facing any nation ill prepared for public health disasters: "Now that we are less than a day by plane from the jungle-type yellow fever of South America, less than two days from the sleeping sickness of equatorial Africa, less than three days from cholera and bubonic plague, the ramparts we watch must be civilian in addition to military."

Researchers, many of them at the country's most prestigious universities, with new federal research dollars, made new discoveries about the inner workings of genes. First, it was discovered that the method used by genes for exchanging genetic materials was achieved through a large molecule made up of sugar, bases, and phosphoric acid, known as deoxyribonucleic acid, or DNA. By the 1950s, the hottest story in the world of science consuming scientists across the globe was the race to uncover the molecular structure of DNA. Understanding its structure was the first step toward understanding how DNA worked and thus the key to human life.

When the double helix, the twisted-ladder structure of DNA, was finally uncovered with much fanfare in 1953 by scientists James Watson, Francis Crick, and Rosalind Franklin, it set the world of science on fire.

Reading by flashlight in the university library stacks late at night, Ron had stumbled upon the building blocks of human life. And he was hooked.

Meanwhile, what no one realized, not even Ron himself at the time, was that during all those months spent trapped sick on the couch, unable to go to school or play in the woods, he had begun developing unique skills that would enable him to use his brain in strange and new mind-bending ways. Ron had learned skills that would make him a great scientist one day.

In the evenings, he'd join the family to listen to the old-time radio shows like *Fibber McGee and Molly, The Green Hornet,* and *The Lone Ranger.* But during the long daylight hours, when he was lonely with nothing to do, he figured out that if he concentrated really hard he could create far more interesting worlds to escape to deep inside his brain.

"I could literally become any size or shape I wanted," he said. "I could make myself an ant and crawl through the outlet and explore the copper circuits inside the walls. It required me to create images in 3-D so that I could rotate things in my brain. But anything spatial was always very easy for me. It defied the laws of physics. I'd love to get into that state. I'm sure it was some kind of self-hypnosis or meditation because it would really feel real. I didn't realize I was actually developing these spatial skills that would help me someday as a scientist."

When the family bought one of the new black-and-white TVs, he started waking up on his own at six a.m. to watch educational programs on chemistry and physics. And when he got pocket money from working for his dad, he spent it on chemistry and physics books, staying up late at night reading, completely absorbed by the sciences. Since math was helpful in understanding this new world of chemistry, he asked his sister to teach him all the math she was learning in school. Patty, who didn't want her younger brother to be stupid forever, got out a chalkboard one day and willingly began to oblige.

When Ron entered high school, his reputation as a low achiever had already been cemented. His guidance counselor told him he wasn't college material and signed him up for shop class. He said OK, but he also wanted to take algebra. When the counselor said no, he wasn't qualified, Ron showed up to class anyway, eventually forcing the counselor to register him for the class. The same thing happened the next year with chemistry. And it was there he began to really shine. Remembering molecular structures was a breeze by then, and his chemistry teacher didn't care that he didn't know how to spell.

In 1960, when it was time to graduate from high school, it was Ron's biology teacher, Charles Compton, who, along with his chemistry teacher, told him he had to go to college.

"I was not even thinking about going to college," Ron said. "I was thinking about trying to get a job at a chemical factory. The guidance counselor said he wouldn't help me get into college or get a scholarship because I was a waste of his time. But based on my vocational and other test scores I had the aptitude to be a carpenter or a mortician. My English teacher sympathetically told me, 'Oh no! You shouldn't go to college. You'll never make it.' But Charles Compton told me, no no, you've got to go to college. Fortunately, the town had a college, and I knew the library there quite well," he said chuckling. By then, Ron had been sneaking into the college library at night for years. Patty, who had always thought of her little brother as something of a wuss, admired her older brother, whom she described as being like "James Dean in the movie *Rebel Without a Cause.*" Who knew Ronnie was the real rebel after all?

Ron's dad was totally opposed to the idea of college, but he said he wouldn't stand in his son's way. He would allow Ron to keep living at home for free and attend the nearby Eastern Illinois University, but Ron had to pay the tuition and fees. So Ron found a way to rent textbooks to save money and quickly got a job on the college campus working in the chemistry department's stockroom. "Of course, I explored every chemical there," he remembers all these years later with unrestrained glee. Next, he proceeded to take every math and science class available on campus in order to balance out his poor English grades and keep a high grade point average. He ended up qualifying for degrees in math, chemistry, physics, and botany.

"It was my physical chemistry teacher in college who said, 'You have to go to graduate school,'" Ron said. "I assumed he meant the University of Illinois in Urbana, but no, he said, 'You need to set your sights higher. You need to go to Caltech or Stanford or Berkeley.' I said, 'What? You think I can get into those schools?'" Ron had dreams of moving out West, far away from his small town. It was the sixties. Out West, he imagined, there were all kinds of new worlds to learn about. He wanted to grow his hair long and meet a California girl. So he shrugged and went ahead and took the Graduate Record Examination (GRE) required for the application. When he got the scores back, he scored in the 99.9th percentile on the science and math portion of the test (Ron is still arguing about that one question he missed). In the English section, he scored in the bottom 17th percentile.

Next, since applications to graduate schools were free back then, he sent his off to each of the three schools out West, along with the one in Illinois, and waited for rejections. When he heard he'd been accepted at all four, he was "blown away." He'd read about a Caltech professor named Linus Pauling who'd won the Nobel Prize in chemistry. That impressed him, so he chose Caltech, got on a plane for sunny Pasadena, and never looked back.

Chapter 6

Blood Trails

The Stanford Genome Technology Center is a modern office building with long fluorescent-lit hallways and a wide-open lounge space on the first floor, where Ron's office is also located. Underneath this spacious lobby, in the basement is the "wet lab," rooms where the science experiments get conducted under overhead hoods filled with expensive lab equipment like mass spectrometers, high-speed centrifuges, and high-tech microscopes along with many custom-built instruments unique to the lab.

It was first built for a now-failed Silicon Valley startup, Ron told me once when I asked why it wasn't like the other science labs I'd visited on campus, which tended to be in older, darker buildings with lab equipment crammed into every open space. This building was spacious and had a high-tech feel. Ron told me the lab was first set up in a different building across town two decades after he got his PhD at Caltech. By then, Ron was well established as a tenured professor at Stanford and reaping the rewards of monumental discoveries made over his career in the field of genetics. Terms I didn't yet understand like "sticky ends" and "polymorphisms" linked Ron's name to the eventual launching of the Human Genome Project. He became codirector of the Genome Technology Center in 1989 and a few years later became director, remaining at its helm ever since, mentoring reams of PhD students and other young scientists as they've passed through the lab. It was his current cohort of budding scientists that Ron turned to in order to build an ME/CFS research team.

His first recruit was Laurel Crosby, an engineering research associate who knew Whitney back then as "the boss's son" and the guy who did photography projects at the lab. She had been through a similar situation to that of Ron's family: she had spent two years getting a rare diagnosis for her infant son.

"There is something about the feelings of helplessness that parents feel when their kids are suffering, no matter what their age," she told me. "It wasn't work at that point, just discussions about how to tackle something so baffling. Ron would later call ME/CFS 'the last major disease we know nothing about'—an enigma just waiting to be solved."

I first met Laurel during my research for the *Stanford Medicine* magazine article. Two years later, we crossed paths again in a hallway at the lab when I was there to sit in on one of the ME/CFS research team's weekly strategy meetings. Laurel, whose dark, curly hair showed touches of early gray, stopped to talk.

"Hi Tracie, how are you?" she asked. "I'm glad you're here." She remarked, "This illness is like a jumble of hidden clues, so it's like working on a giant puzzle."

I told her that I'd decided to write a book about Ron's story, that it was my way of helping, to spread the word about this disease, and asked if she had time to sit down and talk for awhile.

On a couch in an empty hallway, we sat together, me sipping coffee. I didn't want to interrupt her work, but I also knew that all researchers love to talk about a project they're passionate about. I'd talked to so many researchers at Stanford about their work and was inspired by the endless hours of tedious hard work they spent alone in their laboratories, sometimes working for years on end trying to solve some basic science question. It could take decades before their discoveries resulted in creating new treatments or curing disease. I was still hoping that potential treatments, even a cure, would come much faster than that for these ME/CFS investigations.

"How did you first get started researching ME/CFS?" I asked.

"One day Ron walked into the lab and handed me a box, asking me to study the specimen inside," she said. "It was a box of Whitney's poop!" We both burst out laughing.

When Laurel first started this investigation, the lab had little money for ME/CFS research, so most of the recruits had to carve out a few free hours from their other, better funded research projects. Because of the lack of funding, all that the small cohort of scientists could really afford at first was to study one patient. Luckily, Whitney had volunteered. He donated not only blood samples but urine, excrement, and saliva to the cause. Ron also planned to study all the microorganisms in his son's gut, referred to as the microbiome.

"We were given a task," Laurel said. "Our task was to get Whitney out of bed."

Laurel, eventually joined by others in the lab, also volunteered to help with on-call pick-up duty when Whitney's blood samples were ready, so that Ron and Janet didn't have to drive over at night. Usually she'd be at work in the lab when she'd get a phone call from one of them in the late afternoon. She'd drive over to Ron's house as quickly as possible, pick up the vials of blood, and speed back.

"When I got there, I would walk around to the back of their house to the screen door, and Ron would hand me a Ziplock bag with one or two tubes of blood," she said. "It takes me about twenty-five minutes to get back to the lab, but worse during Palo Alto rush hour traffic, so I kept to the back roads with the Ziplocked bag tucked upright into my cup holder."

Back at the lab, she walked briskly down the long, white hallways, to the stairway that led to the basement's biohazard lab. There she pulled on a white lab coat and a pair of white plastic gloves and started the processing of the blood. Using a much fancier centrifuge than Ron kept in his toolshed back at home, she watched Whitney's blood spin around as it separated into its separate parts—red blood cells, platelets, and plasma. Then she ran the different components through a saline wash three or four times. After counting each of the cells, she then handed off the suspension to a lab tech to continue the process. They further divided the blood into its even smaller parts—white blood cells and DNA.

"With Whitney having a messed up day-night sleep schedule, it meant that everyone else had to work nights," Laurel said. "Essentially everyone was on standby."

I had always been impressed by Laurel's and the other lab members' dedication to the work, but now I was even more so. Their investigation was motivated by more than scientific passion; it was motivated by compassion for the boss's son.

When the job was complete, usually around seven or eight p.m., some of the blood samples got handed off to other lab recruits, who worked long into the early hours of the morning to get the job done.

While Ron and his team at the lab continued to build on their research findings, doing the best they could with what limited funding they had, Whitney continued to deteriorate. He stopped listening to music on his iPod. Only white noise came through his earphones now. The further Whitney slipped away, the more intense Ron's research became.

Whitney had begun to write notes to his mother on index cards. Janet would cut dozens of them in half and leave short stacks held together with rubber bands on the left side of his bed within arm's reach. When he could no longer write, he spelled out his needs using Scrabble tiles. Then there was almost a two-year period when he could no longer use the Scrabble tiles, and he stopped communicating at all.

Ron began to dream of science experiments at night, then wake up in the morning with a new piece to add to the ever-growing puzzle expanding in his lab. He'd fallen into a routine in which he would come home around four p.m., care for Whitney in the afternoon, hand over the caregiving to Janet for the night, and then go back to work, disappearing into the library with his laptop, lost in the science and hunting for answers.

Back at the lab, Whitney's blood samples were being put through hundreds of thousands of tests, using the advanced technological equipment at hand. When that wasn't enough, Ron sent off samples to other high-tech startups, often run by his former students, for other tests, including one to measure all the metabolites in Whitney's blood. Or his lab would create new testing technologies. By now, the $3 billion cost to map one human genome had fallen into the thousands. Ron was able to get Whitney's genome sequenced three, four times. Each time it came back, they ran it through more analyses, hunting for hidden viruses, bacteria, and genetic mutations. They were looking for abnormalities, defects, and any unusual patterns.

Finally, results began to come in. The scientists started to uncover abnormalities in the cells, including an overactive immune system and unusual gene mutations. Ron was already beginning to formulate his next plan for what he called a "big data" study. He wanted to run similar tests on twenty other severely ill ME/CFS patients like Whitney, as well as some healthy volunteers.

Much of science is based on replication of results and finding trends in larger population samples. Any single result needs to be tested over and over again before it can be trusted. Ron believed the molecular signals for this disease would be strongest in the most severe patients, like Whitney. No one had studied this group of patients before because they were housebound and couldn't travel to medical clinics. No one really even knew where they were. But first, before that next hurdle could be crossed, the team needed more funding. Money from the Gruber Prize had run out. And the family's efforts, spearheaded by Ashley, to raise funds through the Stanford CFS Research Center just weren't bringing in enough.

One day, Janet attended a Bay Area support group meeting for ME/CFS. Whitney had gone several times when he was still healthy enough. She was too busy to go often, but she went when she could to meet people in the community and get ideas for things that might help her son.

She liked to talk about Ron's progress with his research. It always seemed to bring hope to the sick people there, who were often desperate for help. That day, they had a special speaker: Linda Tannenbaum, who had a daughter with ME/CFS. She had started her own nonprofit to help others with the illness. Janet said she was impressed by Linda's fundraising knowledge.

Linda's daughter got sick when she was sixteen years old. She was fine one day, and the next day she was suddenly housebound and missed all of eleventh and twelfth grades. Like Whitney, she went through a litany of medical tests that came back negative, no one knew what was wrong, and yet she struggled to get out of bed. She visited twenty doctors before the last one finally diagnosed her with CFS, but he couldn't help her. She suffered unrelenting pain, migraine headaches, muscle aches, cognitive dysfunction. She had severe orthostatic intolerance, meaning that when she was upright, she often passed out. This led to multiple emergency room visits. She was prescribed physical therapy at one point that made her much worse. Pain and sleep medications offered no relief.

"She got bad enough, many times she just passed out," Linda said. Then, during an extended hospital stay, her doctor summoned Linda and her husband to meet for a private consultation.

"I truly thought he was going to say she was dying of a brain tumor," Linda said. "Instead, the doctor said that we were enabling our daughter's sickness, and he said that the nurses on the floor agreed with him. He told us that nothing was wrong with her and to take her home.

"I almost laughed when he told us that," Linda said. "She'd been sick over nine months in bed, and here he'd seen her for about five minutes. We refused to take her home from the hospital until she stopped passing out." Eventually they heard about a doctor in Incline Village, a small town by Lake Tahoe; he was one of the few experts on CFS. It was Dan Peterson, one of the two doctors at the center of the outbreak of a mystery disease in the 1980s that first put CFS on the map in the US.

Linda and her daughter flew to Tahoe from Los Angeles repeatedly over a period of a year to be treated by Dr. Peterson. The plane trips were tough on her, but Peterson was their last hope. He prescribed some experimental treatments, including an antiviral and antibiotics. Finally, an antibiotic used to clear up bacteria in the colon for unknown reasons helped Linda's daughter get out of bed. Recently turned thirty, she's not completely free of symptoms but is able to work and live somewhat of a normal life.

"I had told my daughter when she was stuck in bed that when she did get up, I would start a nonprofit to help find a way to diagnose others with this awful disease and fund research for treatments," Linda explained.

For help with publicity, Linda tried approaching celebrities who she believed had ME/CFS, like Cher, who has been constantly linked to the disease in the media, but she had no luck. ME/CFS needed a public relations campaign.

After hearing her talk that day, Janet thought maybe Linda could give the family some tips on fundraising, so she invited Linda to come to their home to meet Ron. And

Linda agreed. She immediately realized that Ron could do the research but was struggling to raise the money. "I told him, if he brought me the researchers, I would raise the money for research," she told me over the phone years later, when I called her at home in Los Angeles. She said Ron smiled and gave her a nod of agreement. For a man with several Nobel Prize winners on speed dial, finding researchers to help who could impress potential donors would be no problem. Within a week Ron had assembled a scientific advisory board with three Nobel Prize winners and five members of the National Academy of Sciences. "Everyone I asked said yes, immediately," he told me, clearly pleased.

The obvious next step to help advance the team's research was to apply for some of that small pot of NIH research money. Ron applied twice for a grant to fund his Big Data study and was twice rejected. They said the proposal lacked a hypothesis, Ron told me.

"But good science first requires observations," he said. When both grants were rejected, Ron broke protocol and made a stink about it. In a letter to senior NIH officials he detailed how the reviewers who gave the grant proposals a poor rating were wrong. He wrote that the comments in the rejection letter were so mystifying, "it made me wonder if they had even read the proposal."

The resulting publicity was successful in two ways. First, it caught the attention of more patients and advocates, who were motivated to donate to the Open Medicine

Foundation in order to keep the lab's research going. Second, it also caught the attention of Francis Collins, the director of the NIH. Collins was one of the former directors of the Human Genome Project. He was quite familiar with Ron's past successes, and he took note when the famed geneticist suddenly had two grant proposals rejected and was loudly complaining about it.

"I called Francis to tell him about my son," Ron said. "I sent him photos of Whitney. I told him how important it was to fund research for this serious disease."

About this same time, the committee that Ron had been working with for the previous two years had finally published its report on ME/CFS for the US Institute of Medicine (now the National Academy of Sciences), a prestigious organization. After reviewing nine thousand previously published scientific studies from around the globe (most of them small and forgotten), the committee of experts had concluded that this was a biological disease with an estimated 800,000 to 2.5 million cases in the US—mostly undiagnosed. The report went on to say that treatments that emphasized exercise could harm patients and should be stopped. This was a biological disease, characterized by symptoms of overwhelming fatigue, unrefreshing sleep, and post-exertional malaise with multiple other varying symptoms, such as cognitive dysfunction. The report made note of the "paucity" of research money and strongly recommended that the government immediately remedy that.

The report made headlines. Ron was quoted in *Science* magazine on February 10, 2015, the day the report was released, saying that he hoped the findings would finally convince physicians that they could, and should, diagnose this disease because it is real.

I thought over what I had learned by now about the history of this disease. I still didn't understand. I knew serious scientists had begun to study it when news of it made headlines in the 1980s, but if it had taken thirty years for a report telling the public and the medical establishment that this was a real disease, to be taken seriously, something had gone horribly wrong.

Chapter 7

Raggedy Ann Syndrome

I joke, but only half joke, that if you show up in an American hospital missing a limb, no one will believe you until they get a CAT scan, MRI, and orthopedic consult.

—Abraham Verghese,
"A Doctor's Touch," TED Talk

IT WAS DEEP INTO winter 2019, three years after my first visit to Whitney's home, when I headed into the Sierra Nevada mountains about a five-hour drive from my home in Santa Cruz. I planned to investigate the strange story of the outbreak in the Lake Tahoe basin of the mystery disease that had occurred in the mid-1980s.

As I drove north through Sacramento into the Sierra Nevada foothills toward ski country, I thought over the old newspaper articles and TV news shows, dating from the late 1980s and 1990s, that I'd researched during the intervening years since I had started meeting with Ron and his family. A media blitz during that time drew the nation's attention to what would become known as ME/CFS. (Since no one knew what this sudden illness was, where it came from, or what caused it, the illness had no name at the time.)

The story goes that two doctors in the town of Incline Village along the north shore of Lake Tahoe reported several hundred patients with an unknown, chronic, flu-like illness to the Centers for Disease Control and Prevention (CDC), the federal agency responsible for controlling the spread of disease. Two junior investigators flew into town from Atlanta, took some blood samples, and two weeks later flew back to the CDC. And then, well, the shit hit the fan.

Journalists got wind of the federal investigators' visit, and overnight the first story blared: "Mystery Sickness Hits Tahoe." The Tahoe patients and their two physicians soon appeared on TV news shows talking about a fatigue so pervasive that previously healthy people were suddenly forced to quit their jobs, many so weak they could do nothing more than crawl to the bathroom from bed. Some took the reports seriously; others mocked the doctors as "quacks" and called their patients "crazy" or just plain lazy or worse. No one knew for sure whether this was something to set off alarm bells, coming as it did on the heels of the AIDS epidemic, or if it was nothing more than a case of widespread hysteria.

I had tracked down two of those original patients who appeared in several of those 1980s news reports: Janice and Gerald Kennedy. They were both teachers at Tahoe-Truckee High School back then. Now well into their seventies, they had moved away to the tiny town of Weimar, population 209, an hour's drive west of Lake Tahoe. From Weimar, the mountains peak at seven thousand feet at Donner Summit, the same spot where the ill-fated Donner Party had been forced into cannibalism while trapped and starving to death in the winter of 1846. From there, the mountains drop steeply toward the Tahoe basin, ground zero for where this mystery began. I wanted to hear the Kennedy couple's story firsthand, but years of media attention had made them leery when I had telephoned them days earlier. Since I wasn't officially invited to visit, I bought some flowers and a copy of the three-year-old magazine story I'd written about Whitney and Ron to help show my good intentions.

The winter storms had come in quick succession through January and February, leaving little time for snow plows to clear the roads. I had hoped not to travel to Tahoe during bad weather, but it had taken months to set up an interview with Dan Peterson, one of the two original physicians at the center of the controversy. He was still practicing in Incline Village more than three decades later, and he, too, seemed leery of media attention. It took months of weekly phone calls to set a date. Now it was February, and temperatures were predicted to drop to 1 degree Fahrenheit. And the snow would be deep.

As I approached the exit from Highway 80 into Weimar, which was usually well below the snow line, the ponderosa pines turned from green to white, and the roads grew slick. After driving several miles down an icy road that led through the center of town—a storefront and some scattered homes—I turned right, and the ground turned to mud. Mud? I hesitated. Then the road dropped, and I began zigzagging across head-shaking bumps and jagged cracks in the earth, passing six-foot-high piles of brown snow at the outer edges of the road. I thought perhaps the Kennedys had decided to hide out as far as they could from the rest of the world. And that this drive was a test to see just how badly I wanted to tell their story. To be truthful, I wouldn't have passed the test, but there was nowhere to turn around, so I forged ahead.

"I bet you don't get many visitors out here," I said to Janice, who met me at the door when I finally arrived at the Kennedys' white home perched on a hill overlooking snow-covered mountains.

"Not many uninvited ones anyway," she said, looking me in the eyes without a smile.

I handed her the flowers, and she let me inside.

Thirty-four years earlier—just about a year after Whitney was born—on a clear afternoon in April 1985, Janice Kennedy arrived home from her job as a high school English teacher and went directly into the garage to get her cross-country skis, as she did every day. She was a skiing fanatic. She skied in mid-winter through blizzards and in the spring, when the sound of her skis scraping over exposed rocks filled the chilly air. It was late in the season, so she picked out the pair of skis she kept for the rocky courses and just stood there for a long moment staring down at them.

"I can't do this," she thought with sudden surprise. "I just can't do this today." She set the skis back on their rack, then went upstairs and crawled into bed.

She and her husband, Gerry, lived in a two-story home built into the side of a mountain near Truckee, a ski town with views of the Tahoe basin's other lesser-known, pristine blue lake, the smaller Lake Donner. The high school where they worked sat along the main drag that cut through the town of Truckee, lined with A-frame lodges, trendy restaurants, and businesses catering to the ski crowd. Teachers worked hard each winter to keep kids in class when the fresh powder arrived, covering the nearby ski resorts, North Star and Sugar Bowl. The Kennedys lived just a few miles away from the school. That day, Janice switched on her electric blanket and curled up under the covers with her little black cat Tinkerbell tucked beside her.

Janice never again would have the energy to put on cross-country skis. The complete body pain, exhaustion, swollen neck glands, and migraine sent her to bed not only for the rest of that day but for many days, and years, thereafter. She had a growing pain in her side that she'd find out months later was a swollen spleen. At her worst, she was completely bedridden, and as the old newspaper articles said, she was only able to crawl to the bathroom and then back to bed again. That day in April, the weakness and mental fogginess that crept over her hung like a heavy, dark shroud. She told me that it felt as if someone were piling bricks on top of her arms and legs, pinning her down. All she could do was lie there and wait for Gerry to come home.

In the weeks that followed, Janice's local doctor ran routine lab tests and found nothing wrong. She tried another doctor in Truckee, who agreed with the first and said there was nothing wrong with her. Still, her symptoms persisted. Somehow, with Gerry's help, she managed to struggle through the rest of the academic year, taking all her sick leave and feeling the same bone-deep weariness day after day. At school, Gerry followed her upstairs to her English class carrying her purse and papers, which were now too heavy for her to carry. One day that spring, several of her favorite creative writing students, who knew she was sick, gave her two dozen red roses.

"I was so grateful," Janice remembered years later, talking to me in her living room. "I set the roses on the desk in front of me so I could lay my head down and rest without anyone seeing me. The students never knew why I loved those roses so much."

When the school year ended in June, so did Janice's career. And now, Gerry, a shop teacher, was sick too. The onset of the disease came on more slowly for him with a general fatigue, bad colds, headaches, an inability to concentrate on anything, and swollen glands. By school's end, he collapsed as well.

"I felt like I was losing my mind," Gerry said of the brain fog that came along with the sickness. "I couldn't think. I'd read a page of a newspaper and couldn't remember a thing. We didn't know what was wrong, but it was horrible. At one point, we were waiting for them to tell us we had some form of AIDS and that we were going to die. It almost would have been a relief."

The Kennedys were not the only ones at the high school to get sick with this mysterious, severe flu-like illness that academic year. Beginning in the fall, teachers had begun taking sick leave, then trying to return to classes, but still so sick they would take turns sleeping on the couch in the teachers' lounge during breaks. Their doctors had told them they couldn't find anything wrong. When they began to hear rumors of an outbreak of a similar mystery-type illness in Incline Village, they all began making appointments with the two doctors who were treating similar patients in the neighboring town, and drove the twenty miles south from Truckee.

Incline Village is nestled between the shores of Lake Tahoe and the Sierra Nevada mountains, just across the border from California in Nevada. Known as a tourist mecca, the town sees a surge in population in the winter, when it draws alpine skiers to nearby world-renowned ski resorts like Squaw Valley, the site of the 1906 winter Olympics. Then tourists return once again in the summer for water sports and hiking.

The town, built among ponderosa pines, includes two golf courses and a Hyatt Regency Casino Lodge and Hotel with breathtaking vistas of one of the clearest, deepest lakes in the United States—the largest alpine lake in North America. In the winter, snow-dusted rental kayaks line the lakeshore, waiting for summer to arrive. The town's economy relies heavily on its tourist and real estate economies, banking on the ski resorts and the nearby casinos to continually draw crowds. But in the mid-1980s almost overnight, the small town became known for something far different. And it nearly tore the town apart.

It started in the summer of 1984, nine months before Janice Kennedy first got sick. Two young, well-educated Incline Village physicians, Dan Peterson and Paul Cheney, began seeing a strange, new type of patient. The two physicians, who were used to the broken legs and torn ligaments typical of a ski town, began seeing formerly healthy patients suddenly complaining of severe flu-like symptoms: swollen glands, body aches, constant headaches, poor concentration, and a general malaise. They couldn't sleep. They couldn't even think. At first, the two physicians figured it was just a particularly nasty flu outbreak. But when the patients didn't get better, they began to worry.

By spring, when the traditional flu had disappeared, this illness hadn't gone away. The numbers of patients had only multiplied, growing to about 160. (Eventually the numbers would peak at about 260 by 1987.) Still, the two doctors, unlike many other physicians in the Tahoe basin, kept trying to find answers. Cheney, who has a PhD in physics from Duke University, was a scientist at heart and had trained to be a medical researcher. For him this mystery disease posed a fascinating challenge. Peterson, an excellent diagnostician who had worked previously on an American Indian reservation, was also known for his compassion. Both doubled down, the mysterious illness growing into an obsession, as questions multiplied and the number of patients grew larger. No one else would treat them.

In late spring 1985, when Peterson learned about the cluster of fourteen sick teachers at the small Tahoe-Truckee High School, along with a sick girls' basketball team and new incoming patients from a nearby casino, he began to consider calling in government experts for help. He hesitated, though. He knew it wouldn't be a popular decision. The other doctors in town were distancing themselves from these unusual patients. Already the word "hysteria" had begun to pop up.

That summer, Peterson, prodded by his wife to take a much-needed vacation, got on an airplane and tried to relax. Instead, he spent the airplane ride reading *Annals of Internal Medicine*, continually looking for answers. What he read there almost made him turn around and fly right back home.

Two studies, one by Stephen Straus, MD, a virologist at the National Institutes of Health, the other by James Jones of the National Jewish Hospital and Research Center in Denver, reported that the Epstein-Barr virus, the virus responsible for mononucleosis, was possibly also the cause of a chronic flu-like illness of unknown origin. Mononucleosis, or mono as it's widely known, is an infectious disease that causes severe flu-like symptoms including headache, fatigue, and an alarming weakness.

"I thought, well that's exactly what our patients are acting like," Peterson told me years later. "Like they had a case of mono that just didn't get better." Finally, he thought, here was something concrete. He called Cheney as soon as he could, and they made plans to test their patients for the virus. When results came back positive for EBV in many of their patients, Peterson picked up the phone and made that call to the CDC in Atlanta to ask for help.

It took repeated calls and months of waiting, but finally the CDC sent two junior investigators, one of them Gary Holmes, who flew into town, met a few patients, collected blood samples, and returned to Atlanta after two weeks. They eventually reported no unusual results for EBV in the blood they collected, concluding they didn't know what was causing this reported epidemic and that, in fact, maybe it wasn't an epidemic at all.

Janice and Gerry remember well how their hopes soared when they heard about the CDC investigators coming to town. Finally, they thought, they'd get some answers. Already, Dr. Peterson and Dr. Cheney were telling them they had this new disease called chronic Epstein-Barr virus syndrome. They wanted to know if EBV was really the cause of their sickness and whether there were any treatments or, they hoped, a cure. They met with the CDC investigators in Dr. Peterson's exam room one day that fall, the two young doctors leaning back against the counter where the cotton ball containers and Q-tips were kept, their arms folded across their chests. To the Kennedys they looked "utterly bored."

"Do you have any idea what this might be?" Janice asked them, when they finished the exam.

"Well, it could be anything, even hysteria," Dr. Holmes replied. Janice sighed, and her shoulders slumped as she looked over at Gerry. Feeling utterly let down, the couple left the doctor's office. By then, people in town had already begun to tell them that they weren't sick—they were just crazy.

The CDC investigation might just have faded away, but the *Sacramento Bee* got wind of what was happening in the small town and ran an article about it, headlined "Mystery Sickness Hits Tahoe." It was picked up by the national wires, and the media frenzy began.

The day after the story ran, the phone in the Peterson-Cheney clinic rang off the hook with calls from across the country. Physicians phoned about patients with similar symptoms, patients called freaked out they had the mystery illness. Local casinos called complaining about the bad publicity that would scare away tourists, their bread and butter. The local newspaper the *Tahoe World* quickly followed up with a story: "Truckee Teachers Recount 'Malady,'" mentioning several teachers by name including the Kennedys. Irene Baker, a Truckee High teacher and friend of the Kennedys who was so sick she had to quit working before the school year was out, was quoted as saying: "It's all I can do to go to the doctor, maybe stop at the store, and get back in bed."

Then the national press dropped in, literally.

"One day in town a helicopter dropped out of the sky," recounted Peterson, sporting a blond mustache and the typical ease of a native Californian, decades later. Reporters for the ABC news show *20/20* emerged from the helicopter and went in search of these unusual patients and their two doctors. The show was viewed in millions of American homes, and the Incline Village Chamber of Commerce went ballistic. But that was just the beginning. Coverage soon appeared in the *Washington Post*, the *New York Times*, and the *Los Angeles Times*. Eventually *60 Minutes* weighed in. Tourism took a dive, and panic set in. "It was either, 'Oh, we're all going to catch it,' or 'Those doctors are crazy,'" Peterson said later. "We weren't crazy. This was no hoax."

It was a story to die for, a bizarre illness in a pretty little resort town, and the media couldn't get enough of it. The news flurry would last for several years before it tapered off. Various names cropped up for the mystery illness, from the "Tahoe malady" to the crass but what was most catchy "yuppie flu."

The *Washington Post* published an article in 1987: "Journalists have called it 'the yuppie flu' after the kind of people—young, ambitious, professional achievers, mostly in their thirties and forties and mostly female—who tend to get the condition."

Another article quoted a patient saying she felt like "Raggedy Ann with all the stuffing taken out," and suddenly "Raggedy Ann Syndrome" was added to the name-calling. The more scientific name, chronic Epstein-Barr virus syndrome, didn't take hold until later, when the disease began to grab the attention of some serious-minded scientific researchers.

For Incline Village, though, it was already a most serious topic. The backlash from the community was swift and severe. The "crazy" doctors and their "lazy" patients became the target of blatant mockery if not outright anger by many in the Tahoe basin. An unnamed doctor in town was quoted in the *Los Angeles Times* saying that the two clinicians reporting the disease were perpetrating "a hoax." The chairman of the Incline Village Visitors Bureau was quoted years later in the documentary *I Remember Me*, referring to Peterson and Cheney as those "quack doctors" who were ruining tourism, adding that the disease wasn't even fatal and that it was "mostly with overweight women."

"The community was terrible to us," said Kathleen Olson, a nurse working with Cheney and Peterson in the 1980s who said she received a phone call from Peterson's son's preschool on the day the *Sacramento Bee* story was published, demanding that he come pick up his son before he made the other kids sick. In nearby Truckee, it was no different. The sick teachers faced outright derision.

"In Truckee everybody knew all about it," Gerry told me. "Everybody had opinions about whether we were malingerers or were actually sick." During those early years when the media continually drew attention to them, Gerry was battling symptoms of brain fog and severe fatigue, while Janice was completely disabled. He was struggling to go to work while doing the housework and putting ice packs on the swollen lymph nodes in his wife's neck. In the winter he could barely lift a shovel to clear away the snow. He recalled watching with horror one afternoon in a Raley's drugstore as a local realtor chewed out a dreadfully sick friend of his, who had the same illness, blaming her for scaring away business. After that, Gerry started shopping at two a.m.

"I just wanted to be able to go shopping in peace," he said.

The Kennedys left their Truckee home, with the sweeping vistas from the living room window and the cross-country skiing steps outside their door, and crossed the mountain to the other side of the Donner summit below the snow line to a new home in Weimar in 1995. They still make the hour-long drive every few months to Incline Village to see Dr. Peterson, but they've managed to find some peace in the rural location, leaving both the snow-shoveling and the constant harassment behind. Peterson never did leave town, and during the thirty or so years since, the harassment has never completely subsided. He continues to live with the reputation as a "quack" from many in town. At the same time, he's managed to treat more than eight thousand ME/CFS patients from around the world.

"We just wanted to start over," Gerry told me after the couple had welcomed me inside their home. Janice had taken my coat and asked me if I'd like a cup of tea. The home was comfortable and tidy with a frosty but well-tended garden visible through a sliding glass door that led to the backyard. She brought me green tea in a china cup, and Gerry smiled. The couple, married for more than fifty years now, worked in tandem, much like Janet and Ron. They glanced at each other, nodding in silent agreement, as they situated me on a couch. I thought of my own husband and smiled at how we cleared out the dishwasher together.

"I'm sorry, but we've had some bad experiences with the media in the past," Gerry said, somewhat sheepishly. Then he boasted about his interview on a *60 Minutes* episode. Not all of it was bad. They showed concern about the road I'd taken to get there. Apparently GPS had led me astray.

Both were gray-haired and pleasant, trim and dressed in similar green-print, button-down shirts. Gerry's face was ruddy and kind. He stood tall. Janice kept her chin tipped high and carried herself in an elegant, if slow, manner to the kitchen for tea. They looked the typical retired couple, comfortable and content. Even healthy. But they weren't. Like so many others with ME/CFS, they looked too healthy to be sick and were repeatedly disbelieved.

But regarding Janice, who has been the sicker of the two, I recognized the tension lines in her forehead that I've seen in other chronically ill patients. I link it to months and years of intense concentration to control pain. Or maybe she was just concentrating hard to make it through our sudden meeting. I knew from other ME/CFS patients that they only had an allotted amount of energy daily or weekly, and I knew she would crash right after. I felt that familiar guilt tugging at me with every minute I spent in their home.

Looking at them reminded me again of how important love was for the survival of those with ME/CFS. Loneliness was almost a symptom of the disease. To me, theirs was a love story.

"We still had each other," Janice said, staring up at her husband tenderly. Without each other, they said, they wouldn't have survived.

After getting sick in the summer of 1985, Gerry somehow managed to struggle through three more years of work. But then the insomnia, brain fog, and body pains got to be too much, and he, too, was forced to resign. Neither of them has been able to work since then.

Peterson still prescribes both Gerry and Janice weekly infusions of a treatment designed to boost the immune system called IVIG (intravenous immunoglobulin) therapy. It's an experimental treatment that insurance doesn't pay for but which they've been doing for many years. They believe it improves their daily lives, giving them a bit more energy to find some enjoyment in life. In 1989, Janice participated in a clinical trial for an experimental drug called Ampligen, an antiviral used to treat AIDS in its early days, available only to patients in clinical trials. Both believe that it may have saved Janice's life. By the time I reached the Kennedys' home in 2019, there had already been a twenty-year-long attempt by the company that makes the drug, Hemispherx Biopharma, to get approval for Ampligen for the treatment of ME/CFS from the Food and Drug Administration. (The FDA has determined the drug is not harmful but wants results from larger clinical trials.)

The couple had fought another twenty-year battle, with Dr. Peterson's help, to get workers' compensation insurance, which ultimately failed. But still, they've been lucky financially. They know others like them who have lost their homes. Before getting sick, they had taken out a disability insurance policy for California teachers. That helped, and they have been getting monthly early retirement checks from the school district as well. So many others have no insurance and no money. I wonder about what happens to the others, not as lucky as they are. Homelessness. Suicide.

"I don't think I can express how much we depended on Dr. Peterson," Janice said. "He kept our spirits up. He never gave up. He was going to find a treatment. I feel so bad for people who don't have that, his reassurance that he would take care of us."

"I know people who are so alone," Gerry said. "I know others put in mental institutions. I know a woman who killed herself. I remember fixing her car once."

Since moving to Weimar, they've learned how to get through each day by living within that "energy envelope" that so many ME/CFS patients, including Laura Hillenbrand and Whitney in his writings, describe. It's the symptom of post-exertional malaise, a hallmark of the illness. You try to guess how much energy your body will allow you to use before it crashes, and you try never to go beyond that line. Or, as Gerry calls it, you "hit the wall."

Gerry's able to garden, at least a bit, growing vegetables and flowers outside. He needs to keep busy for his mental health, and there's not much else that he can do physically, beyond the necessary daily chores to survive.

"This is my therapy," Gerry said, looking out at the garden. Still, sometimes he does too much, goes beyond the envelope and hits that wall. Then he drops his tools where he's working, leaves them on the ground, and crawls on his hands and knees from the garden through the sliding glass door to the living room carpet, where he collapses inside. Sometimes he stays there, sleeping for hours.

Before I left, I thanked the couple for their time and hospitality, and they smiled. Then Gerry glanced at his wife and said, "I guess it's not going to happen in our lifetime, the silver bullet?" I looked over at Janice and raised one eyebrow, asking what he meant.

"A cure," Janice said, and waved good-bye.

During the long drive back to Santa Cruz, I found an immunology 101 course for free on YouTube and plugged my iPhone into the car. I started memorizing the common terms like antibodies and antigens, cytokines and viruses and immunodeficiency diseases. I knew I needed to be better versed in the language.

A few days later, after I got home, I managed to dig up the telephone number for Gary Holmes, one of those CDC investigators who came to Tahoe, and gave him a call. He was living and working in Texas at a military hospital. He told me he had long ago put the aftermath of that visit to Tahoe behind him. All he would say is that he was just trying to do his job at the time.

I thought about what Cheney had told me about those two investigators. He referred to them as "Tweedledum and Tweedledee."

"They were pawns in a great system," he said, describing to me how he felt about those two CDC investigators. "Robots reporting for duty. They grabbed a bunch of blood and left town. That summer we didn't hear anything. Nothing."

Chapter 8

In Search of Lost Time

OVER THE YEARS SINCE Whitney's been sick, Janet's home office, where she once ran her psychology practice, has seemingly transformed into a sanctuary for nostalgia. Just down the hall from the kitchen the small room smells of old books and faded construction paper. Sunlight streams in from the lone window. It's calm and quiet except for the distant sound of a churning washing machine from somewhere deep in the house. Janet and I have tiptoed into the room, which shares a wall with Whitney's bedroom, to dig into the past.

I showed up at their Palo Alto house on my day off to see some of Whitney's early photography and artwork from college. As I researched Whitney's disease, more questions than answers began to plague me—just like Ron, but mine went beyond the science. As a journalist, I found intrigue in the story. In my mind I had begun to intertwine the story of Whitney's happy childhood with the emergence of this mystery disease in the US. I began to imagine this nightmare disease as something evil slithering its way toward this beautiful baby boy, with nobody there to do anything to stop it. Ron didn't even know it existed back then. He was busy cracking other codes as the nation geared up to launch the Human Genome Project. Perhaps it was just because that's the way I interpreted my own life: by milestones in my own children's lives. But today, as Janet tells me more about Whitney's childhood, my mind keeps jumping back to the scary fairy tale I've created of the evil snake that no one ever stopped.

Janet begins to search through the myriad piles of buried treasure hidden in the corners of the magical room, her face begins to glow, and, for a few hours at least, she escapes into memories of happier times.

"I'm a sentimental person," she whispers to me, smiling as she picks up what she says is a giant black widow spider cut and folded out of construction paper. I remember a video of Whitney carving a pumpkin and holding up the spider, fresh and black with a red spot on its back. A black widow. Next she holds up a simple crayon drawing, which she instantly recognizes as a cheetah made by Whitney when he was just three years old. "This is one of the first things he ever drew," she says tenderly.

One art project after another of Whitney's gets held up for admiration. Janet's found the pile of Whitney's kindergarten projects she was searching for, remembering how kindergarten was a particularly productive period in Whitney's budding art career. Whitney was energetic and happy. And he loved art. He drew panthers and leopards and an underwater scene with a very large octopus at its center. There's a drawing of, well, it's hard to say. Maybe a boat? Titled "For Daddy." And a painting of a dinosaur eating a cactus with a large tongue dripping bucketloads of blood. Yup, something my son would definitely have drawn, I think, chuckling. I think back to how much I loved my own kids' art projects, and made sure to get assigned to the "art station" when I volunteered for their kindergarten classes.

"I love it in here," Janet says with a sigh as she plops into a lounge chair and flips through piles of books next to her on the floor. "I used to sit here and read all the time. Now I just store things in here. Whitney can't stand it when I'm in here now. He can hear almost any noise from here, and it's painful for him."

Janet has been hesitant to reminisce much with me about Whitney's childhood. I didn't understand it at first. I'd ask her about what Whitney was like as a baby or in grade school. She'd let out bits and pieces here and there. He loved his sheepdog Hozho, and he loved baseball. But on this day, we are looking through some of Whitney's photography from college, and I guess it makes her sentimental. As I watch her go through all the childhood memorabilia, I remember how, when I did the same thing with my own children's artwork, it was always bittersweet. No matter how much you love them grown and healthy and strong, you grieve for the joy those small children brought you. Then I realize how much harder this must be for Janet. It's not easy for a mother to relive the happy years spent raising two beautiful children while her grown son lies trapped in his bed just on the other side of the wall. The nostalgia turns into grief far too easily.

Janet's old desk is cluttered with framed baby photos of Whitney and Ashley; a wedding photo of her and Ron, when they were younger than their kids are now; seashells; pretty rocks—she collects rocks; and some dream catchers made by the kids. An entire wall of bookshelves are filled with clinical psychology texts and tales of Native Americans and their history. But the floor is where the piles of mementos from Ashley and Whitney's early school days are stacked, and it's here she turns her attention. She allows herself to slip back into those days when Whitney was healthy and it was easy to laugh.

"We always managed to put family first when the kids were growing up," she says. She read to the kids each night. Ron tutored them in science and math. He would sometimes play hooky from work, and the family would spend the day creating tiny new worlds all over the house with little figurines and imaginary landscapes. When the kids grew older, they took violin and piano lessons, and the family traveled yearly to Ashland, Oregon, for the Shakespeare festival, where five-year-old Whitney was captivated by the sword fights.

"When we'd play catch after school, Ron would pitch to him," Janet says with a giggle as if it were yesterday. "I would throw flies like a girl, and Whitney would make fun of me. He taught me to throw and then loved to dive for my fly balls. He got really good at it."

The family took regular hiking and backpacking trips into the High Sierra and yearly trips to a Native American sun dance ceremony in Oregon, a sacred ten-day ceremony with drumming, singing, sweat lodges and constant prayer during which sun dancers fast without food or water for four days, dancing in the hot sun. Ron and Janet not only taught their children a love of nature, music, and science but raised them in a Native American spirituality, something they bonded over from experiences in both their own childhoods. Road trips to American Indian reservations in the Southwest were routine, plus trips to more exotic places like the Galapagos Islands when Whitney was about twelve years old and Ron gave him his first digital camera. I smile with her at their memories. The kids used to play a game they made up and eventually called "Ron.com" to keep from getting bored in the car. They'd ask their dad any kind of question. How deep is that lake? How high is that mountain? Ron would pause for a moment, do some quick mental calculations, then come up with the right answer. They were certain their dad was the smartest man alive.

After an hour or two of sifting through old memories sparked by a handmade Mother's Day card or Whitney's kindergarten handprint, the light in Janet's eyes and the smile on her face begin to fade. The pile of art projects has thinned. And the distant thumping of the washing machine has disappeared. Time begins to move again, as the present day slowly slithers back through the gap under the closed office door. And the light from the window dims.

Janet holds up one last crayon drawing of a shape that looks like a toddler's scribbling and remembers back to when three-year-old Whitney told her about it: "This is a map in case you get lost in the mountains," he announced to his mom. Janet shakes her head, smiling again. Whitney, the adventurer, even loved maps when he was a kid.

"This is making me sad," Janet says, standing up with a sigh.

When I drive home, I arrive after my husband is already in bed. I say good night in the dark and then turn on the light in my new office. I push aside the stuffed unicorn on my daughter's white desk to set up my laptop and do more research. Tonight I'm trying to piece together my own version of those early events—at least, enough to help me understand why, thirty years after the Tahoe outbreak, Whitney had to live through the same horror story that many of those early patients continue to live through today.

Both my children are in college, my daughter, Kaily, in a graduate program at UC Irvine studying poetry. My son, Ben, is playing basketball at UC Santa Cruz, close to home, so we see him on weekends when he needs an extra meal. We see him often.

I grew up not far from Berkeley in the East Bay and met my husband in college, and the two of us managed our lives well enough to be able to escape the crowds of the Bay Area to live by the beach in Santa Cruz. Jogging on the beach became a way of life. He left a law career to teach high school and coach football, and I took a job at the small town newspaper. He was a graduate student in English when I met him, an athlete who wrote me poetry and stole my heart. We had two children, an athlete and a poet, perfect symmetry, I thought. We didn't have much money, but we loved our jobs and each other. It was one of those idyllic pictures from the outside, but if you look too close, the picture looks more like a puzzle. A constant struggle to hold the pieces together. I had managed to keep a writing career going while raising children, but it had taken until now, when their bedrooms stood empty, for me to have enough time for my own obsessions to take hold—a place to put that overactive mind of mine to work. Whitney and Ron had become the latest obsession that kept me up nights.

I picked the story back up in 1987, after the initial outbreak in Tahoe. By then, as Paul Cheney had told me, he had started to get death threats and finally had enough of Incline Village. He wanted to get as far away as possible from the maelstrom enveloping Tahoe. He and Peterson had continued to do research. They had even been joined by a Harvard researcher, Anthony Komaroff, who had begun flying into Tahoe once a month to conduct his own research on their patients and collaborate with them. But not even the temporary glow from Harvard seemed to help tamp down the abuse.

"Dan and I would sit down at a local restaurant, and we would not be served," he said. "Don't call attention to a town that depends on tourism. They blame you."

"Were you scared?" I asked him.

He answered back fast: "Damn right, I was scared."

Cheney responded to a recruitment letter for a job in North Carolina. It would be like coming home for him, a former Duke University Tar Heel. But moving across the country didn't stop his research, and eventually thousands of ME/CFS patients would navigate their way to North Carolina, seeking out one of the few ME/CFS experts in the world. The disease changed the course of his life, but still he often feels like an outcast. "It's a truism for humans: those who get too far from the mainstream get cut off. Even though they may be right."

I found myself wanting to hear more from some of those other early researchers, like Komaroff or Nancy Klimas, a University of Miami doctor and scientist who treated similar mystery patients in the 1980s. Straus, the author of one of the Epstein-Barr studies, was dead, so I couldn't talk to him, although I would have liked to. And I hadn't yet managed to connect with Nancy Klimas: she wasn't at the University of Miami anymore. But Komaroff was still at Harvard and still researching ME/CFS. I'd read that he was one of those doctors who believed his patients when they got sick with a mystery disease and he dove right in to start researching it.

"The early skepticism about this strange disease, instead of discouraging me, encouraged me," Komaroff told me when I reached him by phone. "I was determined to figure it out." I imagined him as Hillary Johnson had once written about him in an early *Rolling Stone* article: with black hair; compassionate, intellectually rigorous; seated in a roomy office in front of a large, dark wooden desk still at his hospital. But of course he'd probably have gray hair by now.

In the 1980s, Komaroff was the chief of general medicine at Brigham and Women's Hospital, a teaching hospital affiliated with Harvard, a position that brought with it a degree of prestige. He had been seeing several of his longtime patients suddenly come down with flu-like symptoms that weren't the flu and wouldn't go away. The patients were extremely ill, and it never crossed his mind that this could be a psychosomatic disease or something just made up. When attempts to treat them failed, and they kept coming back sicker than ever, he started investigating their blood samples in his lab. His colleagues at Harvard showed little interest when he approached them about researching this odd disease, but Komaroff is a bulldog. When he gets ahold of a bone, he doesn't let go. When he heard about the Tahoe outbreak, he reached out to Cheney and Peterson.

"Why do you think so many physicians, and even government scientists, started to believe this was a psychiatric disease?" I asked him.

"I think there is a tendency of human beings, doctors and biomedical investigators especially, when they don't understand something, and they can't figure it out, and it's their job to do so, they blame the victim." He went on. "I think that happens all the time in regards to many medical problems."

By now I'd been carrying around the book *Osler's Web* with me for awhile—the book that Janet suggested I read back when we first met. I bought a used paperback version of the seven-hundred-page book. When its spine broke in half, I carried around the two halves separately. I dropped the first half in the bath once, and it warped. It's full of dog-eared pages now, yellow highlights and notes in the margins. It took some time to read. But I was captured by the story. Hillary Johnson first wrote articles for *Rolling Stone* about contracting this strange disease herself in the early 1980s, when her doctor diagnosed her with something called "chronic mono." I found her website OslersWeb.com, and eventually we set a date for the two of us to talk from her home in New York.

When I got Hillary on the phone, she was both terse and abrupt, explaining that after thirty-three years of being sick, she was pretty "hard-nosed" about it all.

"I got sick in March 1986. I really wish someone had come up to me and shot me in the back of the head.... If the government had just done the right thing, at the right time...." She trailed off.

Osler's Web never made Johnson any money. Not enough copies were sold. There were no royalties. She had gone into debt to write it. She's been too sick for most of her life to be able to continue her full-time writing career.

We discussed the premise of the book. How she had tracked down the few good researchers who were treating the disease at the time. They found evidence that this was a biological disease that affected the brain and the immune system. Most likely it was a virus, they thought. Those early researchers included the two Incline Village doctors, Peterson and Cheney, and a few others like Komaroff and Klimas. The government, though, led by Stephen Straus—the researcher who first touted the EBV causation theory—ultimately made a terrible mistake, Johnson said. "Straus did an about-face and declared the disease a psychiatric malady that occurred mainly in women," she said. "As the primary expert on ME/CFS at the NIH, he influenced grant review boards. Research funding dwindled almost to a stop." At one point, the CDC secretly began diverting funds directed by Congress to be spent on CFS to research other diseases. The scandal made headlines but didn't seem to change much.

"What's the worst thing you can call someone in this society?" Hillary asked me. "That they are crazy. It's a blanket dismissal. It happened to all patients with the disease, and a lot of the scientists who invested serious time in researching it."

I asked about Straus's role and what had happened to him after the end of her book, which covered the subject until about 1995. She told me he continued on as spokesperson for the NIH until he died in 2007. He abandoned his EBV theory early on, when his own research seemed to prove it wrong, and then joined others, including much of the psychiatric community in the United Kingdom, in promoting psychological causation theories until his death from brain cancer. He was the government's prime spokesperson, and he traveled the world spreading the psychological theory.

"He died in 2007 of a brain tumor," she said. "It was one of the best things to happen to ME/CFS research in years. He could no longer play the role of national expert." She pointed out that many people think Lake Tahoe was ground zero for the start of this disease, but that wasn't true. People were getting sick all over—in coastal cities like Boston and New York and many little towns in between. As for her own life, she continues as a journalist to write about ME/CFS when she's able and is happy to finally see a few good researchers like Ron Davis enter the field.

"The government needs to save face now," she said.

Later, I returned to my research and found an old video recording on YouTube from 1996 of Hillary being interviewed about her just-published book on *Primetime Live.*

"Employees at the CDC would make jokes about this disease," she said. "If anyone ever said anything like, 'Gee, I'm tired,' they would be teased about having this fake, bogus disease.... Even the name the government gave the disease, chronic fatigue syndrome, was belittling.... It makes it sound like a benign condition. This isn't about being tired—this is a very serious brain illness. It's an epidemic hiding in plain sight."

Dr. Nancy Snyderman, the ABC News reporter covering the story, addressed the viewers: "We asked to speak to the investigators at the National Institutes of Health and the Centers for Disease Control about chronic fatigue syndrome, but they refused. Instead Dr. William Reeves, the man in charge of investigating chronic fatigue syndrome for the CDC, told us over the phone, that one, there is no viral cause for this problem, two, there are no immune system abnormalities, and three, there are no clusters. When asked about the disease at Lake Tahoe, he said, 'That was hysteria.'"

"I think it's one of the most incredible medical stories of our century, and it's going to be very, very hard for the government to change its position on this disease," said Hillary, brow furrowed, "to have to sort of call up the American public and say, 'Hey, that disease we've been calling chronic fatigue for ten years, it's really something far more serious and it's transmissible.... We made a mistake in Tahoe, and we've been making it ever since.'"

The year that *Osler's Web* was published, another significant event occurred that would exert far more influence on the future of ME/CFS at the time than her book would. The British Royal Colleges of Physicians, Psychiatrists and General Practitioners published a report that got a lot of attention from US federal health agencies.

I read a copy of the report online. It said that personality factors and psychological distress were the primary cause of myalgic encephalomyelitis—which was by then equated with CFS in the US. It espoused two treatments. The first was something called graded exercise—which means slowly increasing exercise levels—and the second was cognitive behavioral therapy. In other words, get exercise, and go see a mental health therapist. The report also said that there was a general misunderstanding by doctors and their ME/CFS patients that physical activity should be avoided. This was something that needed to change.

The report outraged the patient community, as it still does today. Advocates call it "one of the most damaging documents ever published for the treatment of patients." But US health agencies, led by Straus, didn't agree. That same year, the CDC published its own treatment guidelines for CFS, recommending a slowly increased exercise regimen and behavioral therapy. The recommendations closely paralleled the UK treatment guidelines that are still in use today.

In 1997, Ron knew nothing about Hillary Johnson or Tony Komaroff. He'd barely even heard of chronic fatigue syndrome. He was hard at work in the lab but took some time off to plan a backpacking trip to Mt. Whitney that year to celebrate his son's upcoming fourteenth birthday. It would be just the two of them.

They took off late in the summer, just before Whitney was to start his first year at high school, their goal to make the seventeen-mile trek to the top of Mt. Whitney, the tallest mountain in the lower forty-eight states, which peaks at an altitude of almost fifteen thousand feet. In the spring and early summer, ice axes and crampons are needed, but by late summer, when the two set off, there was no snow or ice, just the thin, bright air of the high altitudes.

Even though it was summer, at that altitude it was cold and barren, and the rocky trail filled with boulders was a strenuous climb. Whitney's parents named their son after the mountain because of their deep connection to the wilderness in the western United States, where they'd found the beauty and freedom to live the lives they both loved. This was meant to be a coming-of-age trip for Whitney, as he started the long trek toward becoming a man. It was a difficult climb; at one point he wanted to give up. But his dad encouraged him on, as they crawled across the perilous trail that dropped down thousands of feet on either side. Ron led Whitney through the journey, telling him he would make it. Trust in Dad, and things would be alright.

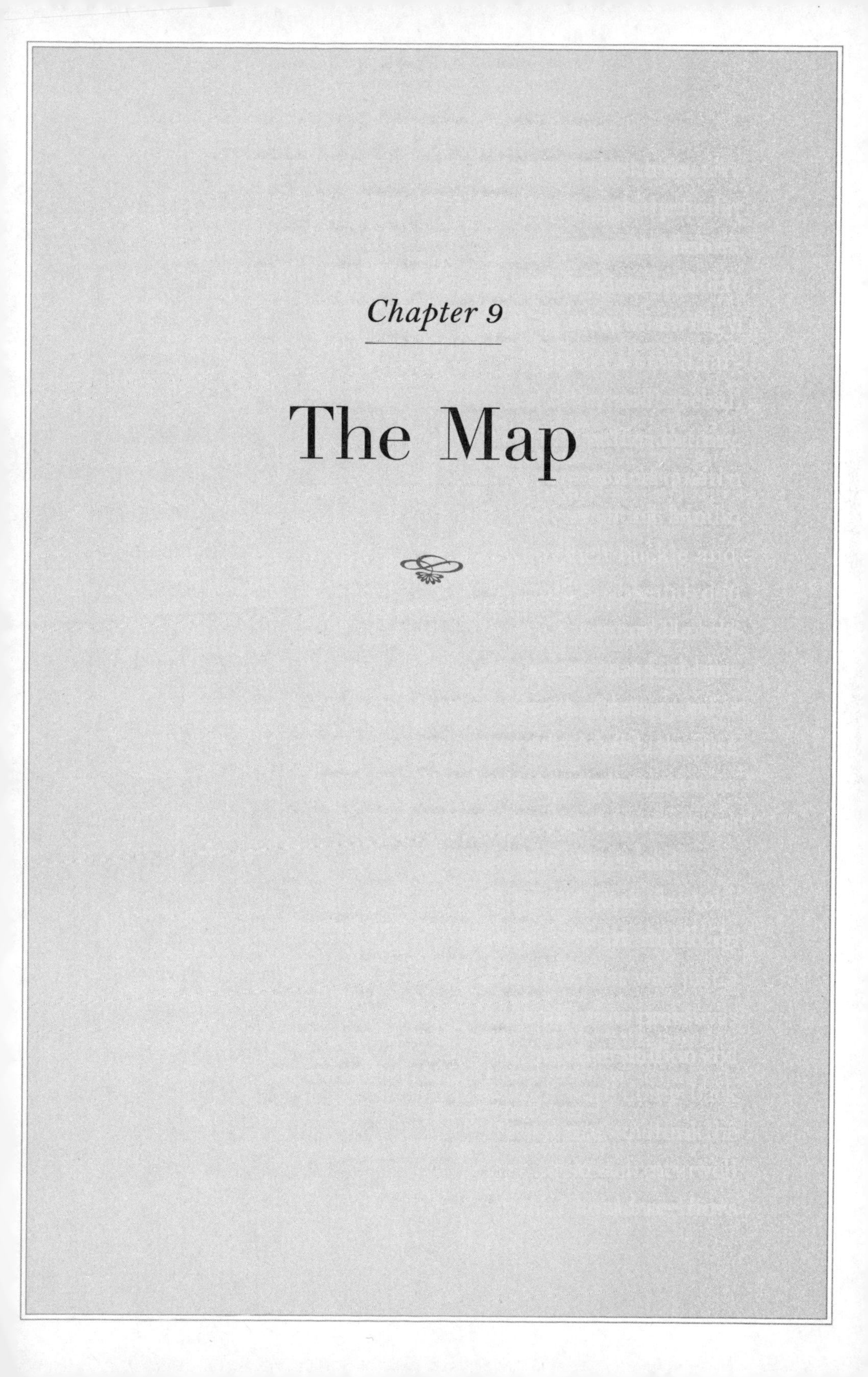

Chapter 9

The Map

RON SHUFFLED INTO THE library from the living room, which has a grand piano and hardwood floors. He was bleary-eyed and shoeless, the image of the absent-minded professor. His white hair was smashed flat by the couch and sticking up at odd angles. It was late afternoon on a Tuesday, and he'd been napping on the living room couch, his jean jacket draped over his shoulders, knees tucked up, and stockinged feet hanging over the edge. He was up late the night before, cleaning out Whitney's commode while Janet hooked up his IV line. The commode has to be cleaned up quickly because smells can make Whitney crash. It made me smile, thinking of my own dad, an engineer, now passed, who loved naps.

I had come over for a visit after work. Janet and I had started talking about Ron's career when she had rummaged around for a certain bunch of old family photographs and spread them across the coffee table, most of them of Ron. She loved to talk about family lore, and she laughed out loud when she showed me a photo of Ron as a child playing the accordion. The three of us met occasionally like this at their home in the library. Ron would try to fill my head with details on how the field of genomics was born, while also catching me up on what was happening with the current ME/CFS research in his lab. That night

I had called them in advance to see if I could pick up dinner. They said sure. I found a nearby pizza parlor, one with white tablecloths, where I got some fancy mushroom pizza to go, at a ridiculous cost. Married to a school teacher with two kids in college, I was on a limited budget, and the price caught me off guard. Palo Alto prices were always a shock to me. The cost of living in this town kept me and most of my coworkers commuting long hours. Few could afford to live here.

Ron yawned, then perked up at the photographs and started shuffling them around. There was a snapshot of Ron as a newlywed laughing with Janet, his young bride; another of him as a pink-cheeked four-year-old posing with his older sister, Patty; and a more recent one of him carving a cradle for his soon to be newborn son. Several more provided a glimpse into Ron's career as a scientist, as he quickly rose through the ranks of academia, from Caltech to a postdoc position at Harvard, then on to the spot he most yearned after: a faculty position at Stanford. There was one of a studious-looking Ron, in a turtleneck, animated in front of a giant blackboard addressing a crowd of the world's leaders in molecular biology at the now famous Asilomar conference in 1975 on the ethics of recombinant DNA. And another of him shaking hands with the pope during a trip to the Vatican to discuss the ethical ramifications of genetic engineering. Cool, I thought. And then, way back at the beginning of his career, as a graduate student with a faint wisp of a mustache, he is reaching up high to adjust that giant electron microscope in the basement at Caltech. Ron picked up this photo and smiled at it.

"Caltech was a miracle for me," he said. "I was scared to death, but I loved it right away. I decided I was going to enjoy myself and learn as much as I could until they discovered I was too stupid to be there and kicked me out. I still didn't know that I had dyslexia." That got Janet talking about how the two of them met at Caltech when she was a freshman at nearby Pomona College.

"He was like this poor country boy with a rope tied in a knot as a belt," she said, and they both guffawed. The nickname Farm Boy stuck. They met at a dorm party thrown by a bunch of physics graduate students—must have been a real rager, I thought, though not unkindly.

"Our first date was going to Dino's for dinner, this Italian place in Pasadena that has the best minestrone soup ever. Then he took me to the Ice House, a club where we heard the Irish Rovers. You know the lyrics 'You're never gonna see a unicorn'? There were velvet pictures on the walls lit with black lights. It was dark. I thought, *Wow, this is cool.*"

For their second date, they went to Disneyland to see the fireworks at night. "He told me he made arrangements for Disneyland to stay open late for us," Janet said.

"And she believed me," Ron added laughing.

Ron started to talk about his career then, skimming over his first famous paper at Caltech and moving on to research at Stanford. Over his career, Ron has had several of what scientists call aha moments, when a puzzle they've been trying to solve suddenly fits together and makes sense. Two of his most famous came when he was conducting research at Stanford, one in a single night just months after he first arrived. His lab was next door to Janet Mertz, a grad student of Paul Berg, professor of biochemistry, whose office was just across the hall. She was working on Berg's pioneering research to cut and paste DNA from different organisms, which would win him the Nobel in 1980 and usher in the era of gene splicing. But his process of doing this was long, complicated, and tedious.

One day when Berg was out of town, Janet came into Ron's office puzzled by an unusual finding. She had cut a circular viral DNA with an enzyme that should have made it linear and dysfunctional, but there it was, a piece functioning like normal.

"I immediately realized that the ends of the DNA had not been cut straight across, like everyone thought, but rather, the enzyme had cut it diagonally, at different points on each strand of the molecule, leaving 'sticky' ends that could be recombined with other pieces that had been cut similarly. This was a simple method for creating recombinant DNA. Now that it was easy to do, many more scientists started doing it," Ron told me. "That really sped things along."

(When I told a coworker of mine who wrote regularly about genetics research that Ron Davis discovered "sticky ends," she grew goggle-eyed.)

I was scribbling faster and faster in my notepad by that point, struggling to figure out the way his brain worked. Sometimes I'd pause and make the notation "lost!" in the margins and hope my iPhone was doing a good job of recording his words. I knew Ron's brain worked differently from most people's, even other scientists. That he'd rather draw spatial diagrams to communicate his ideas than use words and sentences. He'd told me once that sometimes after he'd found the solution to a problem, it might take him a year before he could verbalize it to others.

It had become more and more important to me to understand the science. How else to understand this man or this complicated mystery disease he was now tracking? Sometimes I'd just throw my hands up in the air (figuratively) and find another expert. Since Paul Berg, now in his nineties, happened to be still at work on the Stanford campus, I tracked him down to help me understand sticky ends and Ron Davis, the researcher.

"That was a major breakthrough that Ron made together with my graduate student Janet," Berg told me, the two of us seated in his small office. I'd gotten over my nerves at meeting Nobel Prize winners years before and could now just sit there and admire his clear mind and the excitement and energy he exuded when he talked about science. "They showed the enzyme cut the DNA molecules in a staggered way, and that enzyme became the magic enzyme to make recombinant DNA. But that was typical of Ron. He had an insight for finding tools to solve difficult problems. He is just extremely clever."

Another of Ron's most famous aha moments came later in his career, when he was studying polymorphisms in yeast, which are places on the genome that have mutated so that different organisms have slightly different sequences at those locations—a sort of unique genetic freckle found on individual genomes.

In April 1978, Ron and David Botstein, a geneticist from MIT, traveled to a University of Utah conference held in the ski town of Alta in the Wasatch Mountains. What captured their attention was a graduate student and his adviser, geneticist Mark Skolnick, discussing their attempts to painstakingly map the inheritance of a gene for a specific hereditary illness in humans. But Skolnick was striking out. As Ron and David sat in the audience listening, they began to wonder why the researchers weren't doing it a much simpler way.

"Both David and I had discovered the use of DNA sequence variants or 'polymorphisms' to do genetic mapping in yeast," Ron said. "We'd been doing this independently in our own labs since the early 1970s. I didn't realize scientists were not trying to do this in humans until then."

Sitting at a bar at the end of a long day of meetings, he and David had a discussion with Skolnick about his attempts to map BRCA cancer genes using protein polymorphisms.

"Why are you using proteins?" David asked.

"What else should I be using?" Mark replied.

"DNA!" Ron exclaimed. David and Ron then explained to Mark how it would work. This was the birth of the idea to use polymorphisms in DNA to create genetic markers for identifying and isolating disease genes. It would also lead to the use of DNA in forensic science and eventually the construction of the entire map of the human genome. One day, it would also allow Ron to get his own son's genome sequenced.

And now there was another aha moment, not nearly so famous as the other two but one that was by far the most important to a father who was trying to save his son. It happened much more recently, during the time that he had formed the ME/CFS research team in his lab and was running all kinds of technological tests on Whitney's blood, their only patient sample at the time. They were casting that broad net, as Ron had described it, hoping to catch some fish. And finally they did.

One night, in the spring of 2015, Ron and Janet were together again in their library. Ron was working on his computer when his eyebrows suddenly shot up and he motioned to Janet to come over to look at something on the screen.

"I thought, *Wow*," Ron told me. He'd been waiting for two years to get the results of a test that would show how his son's metabolome was functioning, a chart of all of the metabolic processes going on in his body. I'd seen the chart once before, when Laurel showed it to me on her computer screen in the basement of the lab. She'd been trying to deconstruct Whitney's ME/CFS symptoms into molecular mechanisms. I imagined the chart, which looked like an enormous dot-to-dot puzzle with different chemical reactions leading to endless others in an enormous interconnected mass, endlessly whirling around in three dimensions in Ron's ever-active mind.

He looked at the chart for about three minutes and immediately saw that Whitney's body was out of fuel.

Ron drew his fingers over the lines on the chart, showing Janet evidence of vitamin and nutrient deficiencies caused by Whitney being fed intravenously at the time. Then he pointed out a whole lot more deficiencies in the biochemicals, or metabolites—the amino acids, carbohydrates, lipids, and nucleotides—that were needed to make energy and keep the machinery of the body running. Of the 700 measurable metabolites plotted on the chart, Whitney had abnormal levels of 193.

"I know what's wrong," he told her. "It's the citric acid cycle that makes ATP. That's why he's so fatigued. It's probably a problem in each individual cell." The citric acid cycle is a series of chemical reactions within the cell that use food to generate energy in the form of the molecule adenosine triphosphate (ATP). Every cell uses ATP for energy. It's the fuel that keeps the cell going. It was clear that several metabolites in the citric acid cycle were abnormally low, suggesting that Whitney was having trouble making ATP. Ron was excited because he knew he had the biochemistry skills needed to investigate the problem.

Still, he knew that it would be no easy task. Given that there were 1,600 genes involved in the workings of mitochondria, where ATP is made, finding what was wrong would be a daunting task. Still, in that moment, they both were elated. Here was irrefutable scientific evidence that there were defective molecular pathways in Whitney's body and that ME/CFS was a real disease.

Ron and Janet were in tears.

By that summer, Whitney had been living off of liquid nutrients, delivered through an IV line, for more than a year. Ron kept running more updated metabolome tests that showed Whitney was worsening, that he desperately needed to get more nutrition, to keep his heart going, his blood pumping, and his metabolism running. He now had three doctors: two CFS specialists, including Eric Gordon, and a general practitioner. All three said Whitney needed to get a feeding tube, called a J-tube, to deliver pureed food and other medicines and nutritional supplements that an IV line couldn't deliver. Janet, feeling helpless and desperate, started the hunt for a surgeon, since that was the only way he could get the feeding tube inserted. She eventually found one on a recommendation. On a sunny day in June, she and Ron sat in an exam room waiting for a consultation with a well-known surgeon. Whitney was too sick to come along. Finally they had found a surgeon to help. They were so full of hope. Janet even dressed up and wore one of her Native American jewelry pieces.

When the surgeon arrived, Janet handed him Whitney's thick medical file and started to explain to him that her son's doctors had all said he needed a feeding tube to stay alive.

"He looked at the files for about three minutes," Janet told me.

"I don't think a feeding tube will help," the surgeon said.

"Oh really?" Janet asked, a little confused but still hopeful that he would offer something. "Why? Is there something else that you think could work?"

"No," he said. "Frankly, he needs intensive psychiatric intervention."

This was a statement I'd heard many times during my work in the CFS community. I'd heard it from patients like the Tahoe couple Gerry and Janice Kennedy, who had seen doctors who didn't believe they were sick. I'd been to the same support group that Whitney once had and listened to the stories over and over again. "I can't find a doctor who will treat me." "My medical insurance won't pay for any treatments." "I need someone who will help me."

It was unbelievably shocking to Janet and Ron.

"I can't tell you how upsetting that was," Janet told me. "Ron and I both felt like we'd been punched in the stomach. I'd heard stories like this from other people sick with ME/CFS. I knew it happened. But now I really knew. It was deflating. We really needed help, and here was this arrogant jerk." The surgeon just didn't believe in the disease. It didn't matter how much Ron, a prestigious scientist at the same institution, knew about it, or that Janet had a PhD in psychology. He'd never seen Whitney and only spent a handful of minutes reading their son's long medical charts before making his own diagnosis. He believed Whitney needed to see a psychiatrist.

Chapter 10

The Heartbreaker

ONE SUMMER BACK BEFORE Whitney started getting sick, when he was just nineteen years old, he worked as a photography intern for the biology department at the University of Alaska, Fairbanks. He met a young woman there, Stephanie Land, a twenty-three-year-old aspiring writer living in a cabin without running water and working at a coffee shop in Fairbanks. Years later, Stephanie would write an article about their love affair that summer in an article published by *Longreads*, which I'd happen to stumble upon. But by that time, Whitney himself was far too sick to read her story. He couldn't even look at the photograph of himself that accompanied it.

The two met at a barbecue during that summer in the home of the dean of the biology department, where Whitney was housesitting. He was assigned to work with the biologists at a large animal research station nearby, and he would trail them out into the field to take photographs of their work. Stephanie arrived at the get-together with another guy, but when he left early, she stayed behind.

She fell for Whitney hard, immediately pulled in by his calm presence and the intensity of his blue eyes, a connection as powerful and instantaneous as she'd ever felt before. They both knew that, when the four weeks came to an end, Whitney would get on a plane and fly off. He didn't want to be involved with anyone long term. He wanted to spend his life traveling, he told her. But for the next four weeks, at least, they rarely left each other's side.

They explored the wild Alaskan landscape deep into the brilliant night. They hiked into the mountains, disappeared into the spruce trees, and fed handfuls of lichen to fenced-in caribou. Always, there was music, lyrical and loose, and hot coffee and thick homemade grilled cheese sandwiches. They danced alone together in a yellow kitchen, and the boy took photographs, especially of her. She was his muse. You can hear the folksy, lilting lyrics playing in the background. It's Dire Straits:

Walking in the wild west end
Walking with your wild best friend

The Alaskan summer days were long, and the constant light inspired the young photographer. He'd photograph the light's reflections. How it slid down a spruce tree or caressed her cheek. She watched him watch the world. He'd focus on the smallest details, finding beauty in everything. He taught her to take time to pause and notice the beauty in the mundane. She observed him with awe as he stood mesmerized, watching water from the shower stream over his hands. "Enlightened" is the word she uses most often to describe his spirit, and it surprises her still. Such a serious word for such a young man, especially one who often acted so goofy at the same time.

Mostly, there was lots of laughter and music, lots of music. Whitney lived and breathed music. They listened to folk singers like Bob Dylan and John Prine. He made a music mix for her while he was photographing ice formations out on the Alaskan tundra with the biologists, fighting off the cold and the swarms of bugs. He titled it *Whitney's Cold, Buggy, Work Mix.* Then he hand-scrawled drawings of naked butts on the CD cover just for fun. Still just a kid, she thought.

When Whitney's internship was over, the two traded sweatshirts before he left. She wore his green one from Bennington College, and he wore the red one she bought in Homer, Alaska, with the words "Salty Dawg Saloon" printed across the front.

For several years afterward, they'd occasionally email each other. He'd send her links to folk songs by John Prine with heart emojis. Sometimes they would talk over the phone. He would call to tell her about the small villages in Jamaica where he had lived for a time or what it was like to watch manatees swim beneath a row boat. He seemed very far away.

A year or two later, Stephanie moved to Port Townsend, Washington. She endured bad relationships, became unexpectedly pregnant, and years later had a baby. She lived with her daughter for a time in a homeless shelter, until she saved enough money working long hours as a maid in other people's houses, cleaning toilets and washing other people's dirty sheets. Sometimes she'd think about Whitney, and it helped her through the tough times.

"During my saddest moments, I'd replay scenes with him," she wrote in that *Longreads* article. "In my mind, I returned to Alaska and to Whitney, to the enclosure where I crouched in front of the caribou, and I would remember who I was. Through the years, I fought to maintain the person I had been with Whitney."

I ran across Stephanie Land's article shortly after writing my magazine story about Ron and Whitney. I was scrolling through a few online news sites one night when suddenly I stopped and leaned in to look closer. There, right up close, was a photograph of a young Whitney caught in mid-laugh, lying flat on his stomach on a patch of grass with his eyes squeezed shut. I took a deep breath. Wasn't it lovely? This was the way things should have been, could have been, him healthy, happy, and enjoying life. It was a beautiful story. But it was a mirage. Instead, Whitney's story took a horrible turn, one that ended with me writing about him trapped in a small bedroom struggling for his life. I thought back to him slowly picking up those Scrabble tiles to spell out the word "D Y I N G."

After that, he made his first ambulance ride to El Camino Hospital, where I had interviewed him several times by now. His doctor at the time, Andy Kogelnik, had found a surgeon who would do the operation, and Janet arranged the ambulance ride. Whitney wore his sound-blocking earphones, and his mom wrapped him up in his favorite soft, fuzzy blanket. Dr. Kogelnik, who Whitney had come to love, came to their home to administer Ativan to help him tolerate the ride and the hospital procedure; then he followed along behind the ambulance. Whitney was fragile, but still he kept his eyes wide. For the first time in almost two years, Whitney was going outside. The ambulance drivers pushed him past the hydrangea bushes and the wisteria climbing the sides of the house.

Ashley snapped a photo of Whitney as they lifted him up into the ambulance on the gurney. He was staring up at the wide, blue sky, his jaw dropped open in awe. A sacred moment captured. Then she stored the image on her computer in a file folder filled with many other photographs of Whitney and his father, one with Ron on his knees at his son's side, his hand resting across his heart.

By the time they got to the hospital, something about Whitney had changed. This was the first time he was given Ativan, and it transformed him.

His mother was amazed. "He was suddenly able to communicate through pantomime," she told me. "He was making jokes and smiling. The effects wore off a few hours later, but Whitney could now use the drug about once a month to communicate. He was worried about addiction, so he didn't want it any more than that."

By then, it was the winter of 2016. Whitney's story was getting out to the world. I would interview his parents for the first time and write about it. Whitney was becoming the ME/CFS poster child that Ashley had envisioned. The public relations campaign was underway to help raise money for Ron and his ME/CFS team at the lab. Ashley,

Janet, and Linda Tannenbaum (who founded a nonprofit to help people with the disease) had started to reach out to the media, sharing his story and the photographs. And some coverage had begun to trickle out. That's when the local *Palo Alto Times* ran the story that my editor at *Stanford Medicine* magazine would read, leading me to write another one. I would read the *Washington Post*'s story and see one aired on the BBC. Whitney appeared in a ME/CFS documentary titled *Forgotten Plague*, his condition so fraught it caused the codirector to cry.

The Dafoe-Davis house, by then, had started to draw people to it as well. Janet reached out through social media to others sick and lost, angry and determined to battle this same disease. I sat on the couch in Ron and Janet's library one night and listened to a reporter from Germany interview them for a magazine story. Another day, I'd sit across from a severely ill ME/CFS patient from England, Ben Howell, who was lying on the couch, the trip across the ocean causing him to crash. He told me about how he'd been prescribed graded-exercise therapy in the UK, and it had made him much worse. Ben had connected with Janet over social media and was so inspired by Ron's research efforts that he had risked his health to travel across the world to meet him. He stayed for a month, visiting doctors, hoping to get better treatment than in the UK.

The published photos of Whitney and Ron captured people's hearts, the stories captured their minds, and slowly donations began to dribble in to the Open Medicine Foundation. Most of the money came from patients and their caregivers. Ron now had the funding he needed to conduct his Big Data study. He started to recruit other researchers, like the renowned immunologist Mark Davis at Stanford, a floppy-haired genius who had discovered how immune cells targeted invading infections like bacteria or viruses. Ron needed more people to help out. Others sick like Whitney had begun to put their faith in him, and he couldn't let them down.

Whitney knew little of any of this when it first happened. Most of his life was spent in isolation, both physically and emotionally. His family was unable to tell him about the rejected NIH grants or the successful new funding, or the new researchers who had signed up to join his dad's scientific quest to unravel ME/CFS. He didn't even happen to know who the US president was at the time. That was before he tried Ativan.

More than two years later, after visiting Whitney several times at the hospital, I'd begun to know what to expect. After he got settled into his hospital bed, smiling at the nurses as they checked out his vital signs, he'd ask his mother to leave the room for a bit so we could have some private time. Janet would bristle at his request, loath to lose any of the rare time when she could communicate with her son. But she understood he was a grown man, and our time together seemed to provide him with some kind of a private life.

As months passed, Whitney grew better at pantomiming, and I started to understand more often what he was trying to communicate. Often he wanted me to contact an old girlfriend or friend of his. He hadn't connected with any of them for so long. It was like someone trying to reach out from the grave to touch the lives of those he loved and longed for. It was my job to find them, tell them he was still alive and he loved them still. He figured a reporter should be good at that. It wasn't really my job and didn't help with finding out more information about ME/CFS or even about him. However, each time I'd go home, I'd have an image of Whitney lying in bed, waiting for me to bring him news of his past. Always, I saw my own son's face in his and did my best to comply.

During one of these hospital visits, Whitney seemed more amped up than usual. It seemed that there was a story he really wanted to share or another old friend he really wanted me to find. I knew that sometimes he'd spend hours on end lying in bed memorizing things he wanted to say when he was on Ativan, even the ways he wanted to say them. Then, because his time to communicate was so short, he grew agitated when we didn't understand quickly enough what he was trying to say.

Ashley was in the room with me, sitting on the end of his hospital bed, and we both struggled to untangle what it was Whitney wanted to tell us from his pantomiming.

First, he drew circles in the air, which I knew by now meant he was in India, riding a motorcycle. Then he waggled his head back and forth, looking silly, but quite happy when we guessed he was pantomiming a monk. Then he shook his head. No, this wasn't just any monk, he shook his head harder. This was a special monk who changed his life. Whitney brought his fingertips together to create the tip top of a mountain, and I traveled in my mind back to those Himalayan mountains, where at the early months of his grand adventure he'd had such an amazing time.

As we translated it, Whitney's story began to emerge. He rode his motorcycle into the Himalayas and then traded it in for a backpack, hiking up further into the clouds and finally reaching a tiny town at the tip-top of a mountain. There were six other backpackers with him that day. He held up six fingers until we clearly repeated, "Six hikers." The group sat in a temple at the top of a mountain in a semicircle around this young monk. They stayed there with him for two fingers—two months. I imagined Whitney up there, learning some secrets of life that I'd never understand, when he suddenly threw his hands violently up in the air, filling his cheeks up with air, then blowing out hard. I had no idea what this meant, but Ashley knew. It meant that the monk blew Whitney's mind. Then he spelled out the name of one of his fellow travelers—M A R T A—on his brown blanket. And Ashley said she'd try to find her email in Whitney's phone and connect the two of us.

Whitney's friend was Marta Galligni, an Italian who spent much of her life as a spiritual traveler. Ashley emailed Marta, and I was working at home on my laptop one night when I saw the reply, with two photos attached, and quickly clicked it open to read:

"Hi, last night I read the mail about Whitney and got shocked," Marta wrote. She was living in Italy working as a waitress saving money so that she could return to her spiritual travels in India. Yes, she remembered Whitney—of course. But she hadn't seen him in so long, and she had no idea he was sick.

"Oh my god, I have no words, I have thought of Whitney so much in these years," she wrote. "Can you tell him that I'm trying to get in contact with Shenyen? Shenyen is an English Buddhist monk, he was Whitney's first teacher in Buddhist philosophy. I think he would definitely be perfect for telling you more about Whitney."

I smiled as I read the email, knowing how excited Whitney would be that she had written back. When I clicked open the photograph, the first showed a small group of young hikers lounging on boulders at the top of a mountain with sweeping vistas of villages far below. The second was a small group of backpackers sitting cross-legged on the ground beside a bald headed, fit-looking man wearing an orange tunic. Whitney was in both shots, dressed in the same plain white T-shirt and khaki pants and with a scruffy beard. He looked relaxed and happy, at peace with the world. I would talk to Marta later in Italy during one of her breaks from work. I wanted to hear about this monk that Whitney had grown so animated about, and she was happy to tell me.

"This monk was very special," she told me, in her heavy Italian accent. "He changed my life. I think for Whitney it was the same. Buddhist philosophy is very antique and from the East. This guy, Shenyen, he was able to translate the very old teachings into something that could be understood by us. We were young and not from the East. We were from Italy, and America, Australia and Russia. Shenyen was very clever. He had a scientific approach to Buddhism. I met a lot of teachers, but none like him." She reminisced a bit more about those happy days with Whitney before she had to rush back to waiting tables, but first added, "Whitney, he was such a sweet guy. I loved him. So different from other young kids. I saw that he was searching for something." I asked her one last question before she hung up—mostly it was a question that I wanted answered for myself.

"Do you think Whitney might be using what he learned from this monk to help him live through so many years of pain and isolation?"

Marta quickly answered, "Oh, I would definitely believe so. I believe that Whitney has been practicing all along. Buddhism is only in the mind. His body has left him. He has only his mind." I thanked her and said I would tell Whitney about our talk and share the photographs with him if I could.

"Life is really strange," she said. "Give him my love, big, big love. *Ciao.*"

When I saw Whitney in the hospital again, a few months later, I told him I'd gotten ahold of Marta, and his eyes grew wide. I read the emails and then asked him if he wanted to see the photographs. Usually it was too difficult for Whitney to look at photographs, but this time he nodded yes. When I showed him the photo of his group of travelers at the top of a mountain, a beautiful valley down below, he burst out in sobs. Then he grew angry. He jabbed his index finger at the image of himself, poking at it hard over and over again. That's the real me, he meant. Then he pounded on his own chest, shaking his head no. This is not me. I'm not this pale, sick man, lying in a hospital bed, strapped to monitors, thin and frail. Get me out.

Chapter 11

Vindication

On a gorgeous spring day in 2019, I stood on a street outside the Capitol, the stately dome-shaped home of the US Congress, breathing in the sweet scent of DC's famous cherry blossoms, wondering how exactly it was I had arrived at this spot. The roads were choked with tourists stuck in commute-hour traffic on a Wednesday morning, anxious to see the cherry trees covered with light pink clouds of cotton balls, which also decorated the streets and the wide walkways connecting the Library of Congress to the Capitol.

It was ME/CFS Advocacy Day in the nation's capital, and I was there to get a civics lesson on how to change government. Members of two advocacy groups—Solve ME/CFS Initiative and #MEAction—(those members who were well enough) had traveled to DC to talk to their legislators, raise awareness, and ask for more research funding. There were hundreds of us, wearing purple ME/CFS T-shirts, the majority women, some in wheelchairs, others using canes or wearing masks to protect themselves from germs. Many more looked perfectly healthy when, in fact, they were quite sick. We came from across the nation—from California to Minnesota and Texas to New York.

It was the day before a landmark NIH conference, one that would draw elite researchers, including Ron

Davis, from around the globe to discuss scientific progress made toward finding a cure for the serious biological disease with a bizarre past, ME/CFS. I expected there would be a few researchers from the past as well, looking for vindication. Dan Peterson from Incline Village had told me once that back in the early days, when he first started publishing studies, researchers at the NIH, more inclined to believe the so-called psychogenic theory of ME/CFS, used to pin them up on the wall and throw darts at them.

The morning was clear and sunny when I climbed onto a shuttle with other advocates outside the Bethesda Marriott Hotel, headed to the Capitol, many of us nervous about delivering our three-minute personal stories to legislators on the Hill. Brian Vastag, a former *Washington Post* reporter who has ME/CFS, waited in line with me for the shuttle, and we chatted about the new ME/CFS study at the NIH that he was participating in. We hadn't met before, but I recognized him from his photograph in the *Post* back in 2015, which accompanied an article he'd written titled: "I'm Disabled; Can NIH Spare a Few Dimes?"

The article was actually written as a letter to NIH director Francis Collins, begging him to increase ME/CFS research funding. "Dear Dr. Collins," it began. Previously a communications writer for the NIH, Brian hoped that his story would catch Collins's attention; much like

Ron's rejected NIH grants, it did. "I've long appreciated how the NIH helps the world...," he wrote. "Lately, though, my love for your august institution has been strained. You see, I've been felled by the most forlorn of orphan illnesses."

For me, this day was long in coming. I had traveled far to get here—not just physically but emotionally as well—starting with that first visit with Ron and Janet at their Palo Alto home, and now, here I was in the nation's capital. It was clear to me that just reporting on this American tragedy was no longer enough. It was time to leave the sidelines of objectivity and step out into the fray.

Janet had called me just before I left California. She was frustrated. She wished she could travel to DC as part of the advocacy community. But, like so many other caregivers, she would instead be forced to stay home, to provide care to her son, and to continue to advocate from her computer, reposting on Twitter, Facebook, and other social media sites as news came out from the conference. This was her job: retelling Whitney's story, railing against government missteps, insisting this disease finally get the attention it deserves. "There are thousands and thousands of us who can't come in person and advocate," she said. "You're just going to see the sick folks who mostly look fine."

It was true. Like the young yoga instructor from Washington, DC, who sat next to me on the shuttle ride that morning, who looked fine. She'd come to tell her story to her legislator. She first got sick when she was in her early twenties. Her dad got her in to see doctors at a prestigious research center.

"Those doctors, they told me I was crazy," she said. "All my labs came back normal. It was all like, 'Sorry, go see a shrink.'" Like so many others with this disease, including Whitney, she was anxious for me to know how active she'd been prior to getting sick. She wasn't a slacker or a malingerer. She ran cross-country in high school and studied science in college. She loved to travel. But now she was on Medicaid and had to keep her goals much smaller.

"I just want to wake up not feeling like I've been run over by a truck," she told me.

Tora Huntington, standing next to the two of us in the shuttle, was a middle-aged woman from Massachusetts who had come to speak for her sister, who had been sick with ME/CFS for ten years. Tora asked us if she could practice reading the three-minute personal story she would recount to her legislator later that day.

It was obvious she was nervous. All of us were. I had brought some photos of Whitney to share during my own presentation and planned out what to say as well.

"I'm a retired kindergarten teacher," she said, reading from a notepad in a quavering voice. "My sister is brilliant. She has a PhD in clinical psychology. She worked with children on a Native American reservation and was developing autism programs. Ten years ago, at fifty-two, she was struck suddenly with ME/CFS. We will never get the gifts that she could have given to the world. There is so much living that doesn't get done because of this disease. We are here to raise awareness. To ask for more funding." The despair in her eyes when she looked up left both the yoga instructor and me misty-eyed.

After the shuttle dropped us all off at Capitol Hill, we hiked up First Street Northeast to the government's Cannon House Office Building, where a room had been reserved for the group, since so many of the sick needed a place to rest. It was late morning, and already the room was crowded with exhausted advocates. Some were lying on the floor on top of sleeping bags; others were sipping water and wearing noise-blocking headphones. The room was designed to be a quiet haven for those oversensitive to sound and struggling physically to make it through the day, but the hallways outside reverberated with the click-clacking of high-heeled shoes and the general racket of busy politicians and lobbyists rushing about on a business day. The lack of a reserved shuttle to get back to the hotel at the end of the long day loomed large, overwhelming many who were too exhausted to navigate the crowded subway system in order to get back to their hotel rooms.

My meeting was scheduled with a legislative aide for Congresswoman Jackie Speier, (D-CA Fourteenth District) at two thirty p.m. As I had prepared for the short presentation, my mind had wandered even further back, to just before Whitney was born, and my own journey to become a science writer, a job that now made me proud. I was a premed student at Berkeley who fell in love with anatomy, then grew discouraged by a basic chemistry class. I went on to study mostly public health and stayed an extra year to get a second major in English, burying myself in books the way I had as a child.

By then I knew I wanted to tell stories that helped change people's lives, so I became a reporter and wrote about health and social justice and the things that mattered to me. Years later, when the bottom fell out of the newspaper industry, I was freelancing from home with two small children writing mostly for parenting magazines. I needed a job, and Stanford offered me this one. As unprepared as I was for the job, I worked hard for more than fifteen years to understand science and scientists. Along the way, I came to realize the great value of both in a world filled with disease and loss. A deep love and respect for scientists, and the promise their work holds, now sits solidly within my heart. I've seen how medical scientists can wipe out so much pain and suffering firsthand. I would come to know, all too well, what hell the world could suffer when it refuses to listen to them.

Still, for me, on this day my true heroes remained the patients, those with chronic, ongoing diseases, who somehow manage to get up each morning and go on. And I listened with awe and respect as each one told me their story that day.

Our team met in the hallway just outside Speier's office a few minutes early to discuss our plan. There were three of us: me; Alison Dykens, a young woman who'd gotten sick with ME/CFS as a child; and John Amodio, who was there on behalf of his wife, who had been sick for twenty years. The plan was for John to discuss the problem. Then I'd tell my story about Ron and Whitney, and Alison would tell her personal story. Molly Fishman, Speier's legislative aide, welcomed us in when we opened the door.

John started to talk about this "neglected illness" and how badly a laboratory test was needed to help convince physicians and the public that it is a real disease. It had taken five years and literally dozens of doctor visits before his wife had finally received her diagnosis. Then it was my turn. So I showed photos of an emaciated Whitney lying sick in bed, with Ron kneeling on the floor next to him.

"I'm writing a book about this young man and his father," I told her, pointing to the photos. "He's been sick with ME/CFS for about a decade. He can't eat or speak and has been bedridden for about five years now. His father, a brilliant scientist at Stanford, is trying to find a cure." Molly nodded her head with interest.

"Congresswoman Speier and I are familiar with their story," she said.

Then thirty-four-year-old Alison told her tale. Alison didn't look sick, other than the folded up metal cane in her lap. She was quite beautiful, with glossy, long dark hair. She smiled broadly and spoke in a quiet voice. She got sick with the flu when she was just eleven years old and living with her family in Iowa. Then she started fainting and sleeping up to twenty hours a day. Somehow her mother kept her in school as she struggled through the many years of illness, but Alison never got better. More recently, she had tried working in Los Angeles as a creative manager for Sony, but she had to quit as a result of the ongoing exhaustion. Often she was so worn out that she couldn't do her laundry and would stop at Kmart on the way to work to buy clothes.

Now she lives in a trailer on a bluff overlooking the ocean in Northern California, where she met her fiancé. She didn't think anything this wonderful could ever have happened for her, and she ended her story pleadingly. She told us, "I just want to be able to stand up on my wedding day."

Ron Davis, legendary geneticist, stands before an auditorium on the NIH campus, head bowed to the microphone before him. The room is filled with scientists from around the world studying ME/CFS and, sitting among them, people sick with the disease, who too have traveled from far away to get answers. The crowd grows silent.

"This is an example of a severe patient," Ron says. Behind him, the screen changes from the series of scientific graphs and charts he's been showing to a photograph of Whitney, larger than life, lying on one of his soft brown blankets in the back bedroom of his Palo Alto home. He's wearing nothing but gray shorts and covering his eyes with his hands. He's so thin he looks flattened, as if run over by a steamroller.

"This particular severe patient happens to be my son," Ron continues. "He's covering his eyes because looking at me can cause him great pain.... It always annoys me when people say it's not such a bad disease. This is a very, very bad disease."

Sitting in the audience, I too bow my head. I've seen the photo many times before, and it disturbs me still—not only for how sick Whitney looks but for how exposed he is. Like any mother, I want to cover him up. But he would just push me away. Whitney longed to use photographs to document injustice in the world and would be proud to see his own stark photograph displayed before this distinguished crowd. It's been used by Ron at conferences in Australia, England, Sweden, and elsewhere around the world. Whenever it appears, I imagine the crowd goes silent like this.

Ron begins his lecture on the molecular basis of ME/CFS and his lab's continuing journey to understand it with the Big Data study of severely ill patients. He describes the massive amounts of scientific data collected using the advanced technological tools at hand.

"Severe patients can't even get out of bed," Ron says. "We have to go to them." He tells the audience how, in collaboration with Whitney's former doctor Andy Kogelnik—who made home visits to collect the blood and other bodily fluids of a number of very sick patients like Whitney—they successfully ran advanced lab tests on the urine, blood, and saliva samples from twenty such patients and healthy controls. He expected that the molecular signals would be strongest from these sickest of patients, and he was pleased by the results. "We've been processing and analyzing that data ever since," he says.

Before stepping up to the podium, Ron was introduced at the NIH conference as someone who "really didn't need an introduction" as a leader in the field of ME/CFS research and many other areas of science and technology. The applause was loud, and the crowd seemed to listen hard throughout his speech. He continued on, talking about several of the other studies being conducted in his lab, including one called "the metabolic trap hypothesis" developed by one of his researchers, Robert Phair. He hypothesizes that a mutation discovered in the IDO2 gene may impact the molecular process involved in energy production.

Ron has discussed the complicated theory with me several times. It involves the gene IDO1, which in humans processes the amino acid tryptophan into other compounds that do important things for the body like regulate the immune system, reduce brain inflammation, and produce ATP, the energy juice for the body.

If for some reason there's too much tryptophan in the cell, the IDO1 gene can't process it anymore, which decreases the body's ability to make those other necessary compounds. Fortunately, there is another gene, IDO2, which acts as a backup and keeps the body running as it should. With this two-gene system, all of the necessary compounds continue functioning to keep the immune system healthy and the body producing the necessary ATP.

Unfortunately, about 65 percent of the population has a mutation in that second gene, Ron told me. Most of the time this doesn't matter because the tryptophan levels never get high enough to inhibit IDO1. But with a powerful trigger, like a viral infection, tryptophan can increase and inhibit IDO1. Without the backup gene, tryptophan is not processed into those other important compounds. It's a kind of metabolic trap that can cause extreme fatigue and a dysfunctional immune system. The only way to get out of this trap is to decrease tryptophan levels or somehow reactivate IDO1.

Next at the conference, he skimmed over his lab's metabolomics research, and I thought of how much that first test had meant to him and Janet. He mentions Laurel Crosby's newest study—measuring levels of potentially toxic heavy metals in hair samples taken from patients—and I remember Laurel at a rally held on the steps of San Francisco City Hall, where she was asking for hair samples from the crowd. Nearing the end of his lecture, Ron talked about a study of a potential new diagnostic device, referred to as the nanoneedle. It was developed by Rahim Esfandyarpour, an engineer leading a team in Ron's lab. Ron says that the study had recently been submitted for publication in the *Proceedings of the National Academy of Sciences.* "Today, the journal called me and the study has been accepted for publication," Ron announced. "We should be seeing that out anytime now."

I have to tell my office about the study, I thought. This news needs to get out to the media.

On the morning after Advocacy Day at the Capitol, those of us who were healthy enough had taken shuttle rides from the Marriott to the NIH campus just a mile away—ten miles away from the center of Washington, DC. (Some of the sick folks, just too exhausted by Advocacy Day, stayed behind trying to recuperate in bed in their hotel rooms.)

The campus spread out wide across the hillside, quiet and peaceful, looking much like a college campus, dotted with historic red buildings, blossoming cherry trees, and energetic young researchers briskly walking by wearing backpacks. The NIH is the largest funder of biomedical research in the world, investing nearly $40 billion annually in medical research. Just the fact that a conference on ME/CFS research was being held here, I knew, brought with it a certain level of much-needed credibility to this long-neglected field. As I had walked up the hill to the conference auditorium, I caught a glimpse of the same stately building where FDR in 1940 had broadcast that prophetic speech to the nation on the brink of war: "We cannot be a strong nation unless we are a healthy nation." I figured director Francis Collins must have his office there.

Walking into the conference auditorium felt, to me, like entering the drawing room at the end of an Agatha Christie novel, where all the characters in the book finally gather together to solve the mystery. I didn't expect this mystery to be solved that day, but at least some brilliant researchers had gathered together to try. Over the next two days, in addition to Ron, hotshot researchers from Harvard and Cornell, from Australia and Sweden, talked on and on about their new research findings in the fields of neurology, microbiology, exercise physiology, and more. Low-level inflammation in the brain was reported by one. Another spoke of abnormal anaerobic thresholds as a possible cause for the adverse effects of exercise in patients. During each presentation, the sick folks stayed seated in their chairs, some in wheelchairs, listening hard. They applauded silently by shaking their hands in the air, out of deference for others oversensitive to sound.

At one point, I gasped when I recognized *Osler's Web* author Hillary Johnson walking slowly and painfully up to the microphone set up for audience participation. She had told me she would try to come if she had the strength, but I was still surprised she made the trip. Somehow it made me feel proud to see her there. "People at the NIH may not understand the depths of our despair," she said into the microphone, speaking to the researchers up on stage. "Patients feel this is a public health emergency. AIDS gets $28 billion compared to ME/CFS's $8 million annually. You've got to do better."

Sitting in the back of the room I nodded my head in agreement. We've got to do better, I thought. This story of injustice had continued on far too long. It's time for the misinformation and stigmatization surrounding ME/CFS to stop. Our leaders need to step up to the plate, acknowledge past mistakes, and fix them. All the evidence is there in black and white. More research funding to find a cure would prevent so much unnecessary suffering and save lives.

I spotted the ruddy face of Dan Peterson, who was up front waiting for his turn at the podium. What a day of vindication it must be for him, I thought. When his turn came, he took a moment to reflect on the past. "I'm reminded of coming back here thirty years ago, when I brought a patient for diagnosis and potential treatment," he told the audience. "I saw Dr. Stephen Straus, and he left me with the wisdom that science would prevail. I think that's come true as we've seen in the last two days, but it's a shame it took thirty years."

Throughout the two-day conference Ron wore his dressy black cowboy hat. He likes it when patients can pick him out easily and come talk to him, so he always wears a hat, he said. This could have been an occasion for him to celebrate recent hard-won successes, but for Ron, the only success he longed for—a cure, a treatment for his son—still hadn't come.

After the results from the Big Data study began to roll in, Ron and his ever-growing list of collaborators had shared the data and put it to work, just as he had planned. More studies and more hypotheses began to flow from there. Among them, one by his Stanford colleague Mark Davis—also there to speak at the conference that day—had shown results of an overactivation of those infection-fighting immune cells known as T-cells. This newest evidence of an abnormally functioning immune system needed further investigation. And it was time to submit another NIH grant. I'd asked Mark once what had pushed him into this new field of study. He replied with a smile. "When Ron Davis asks you to do something, you say yes."

In 2018, the year before the conference, Ron had finally got his big ME/CFS NIH grant. The five-year, 2.5-million-dollar project, shared with Mark Davis and another investigator, was designed to find the cause of the overactivated immune system, with Ron leading the gene sequencing arm of the study to help determine what triggered the cells. Meanwhile, fundraising at the Open

Medicine Foundation had reached $9 million in donations, mostly from patients and other advocates, including a $5 million windfall from the Pineapple Fund, an anonymous bitcoin investor. The money had made all this research possible as well as funding the first two scientific meetings and community symposia focused on the molecular basis of ME/CFS ever hosted on the Stanford campus.

Beyond Ron's personal efforts, and perhaps spurred on by them, there had also been signs of a shift within the mainstream scientific community toward a new understanding of ME/CFS. Two years earlier, the NIH had more than doubled funding, from $6 million up to $15 million for ME/CFS—still far below what the advocacy community was asking for, but it was at least a start. When I eventually asked Francis Collins about the trigger for the additional funding, he said it did help that he had been made aware of the much-needed budget increases for the study of ME/CFS through multiple factors. They included Ron going public with the early NIH grant rejections, Brian Vastag's emotional plea in the *Washington Post*, and the publication of the National Academy of Medicine report—the prestigious committee that wrote it just couldn't be ignored.

Finally, in July 2018, the CDC rewrote its treatment guidelines, officially changing the name of this disease from CFS to ME/CFS. Unlike its previous guidelines, it now warned that exercise must be approached with extreme caution for these patients. This came on the heels of what Ron calls a seriously flawed study published by a British research team supporting graded exercise and cognitive behavioral therapy for treatment. It was the same study that Ron and Whitney had discussed with something like shock back when Whitney first moved home and knew how much sicker he got with any kind of exercise.

I followed media reports that had begun to spring up, reporting on these changes. I had begun to note a radical change in the news headlines. "New Recognition for Chronic Fatigue Syndrome," reported the *New York Times*. The journal *Nature* announced: "A Reboot for Chronic Fatigue Syndrome Research." When the nanoneedle study was eventually published, similar headlines would begin to appear.

Before the conference ended, I got a chance to meet Harvard's Tony Komaroff on the way to a dinner party. I'd been invited by Ron to the party sponsored by the Open Medicine Foundation. He too was meeting Tony for the first time. Together with Ron Tompkins, another Harvard researcher and longtime collaborator of Ron's, we took an Uber ride over to the Hyatt Regency in downtown Bethesda, where most of the researchers at the conference were staying. I kept my eyes and ears focused on Tony. It was like a ghost from the past. He did look much like

Hillary Johnson had described him in her *Rolling Stone* articles, with bushy, dark eyebrows—plus the added gray in his hair that I had imagined. I'd heard him speak at the conference, where he was introduced as the "grandfather of ME/CFS research," now the author of more than 150 highly regarded articles on ME/CFS. A dog with a bone that just wouldn't let go, I thought. During the ride to the party, he talked about the need for change in medical insurance to cover the treatment of ME/CFS.

"Doctors just don't get compensated for how much time it takes to treat them," he said. "Things won't change until they do."

When we arrived at the dinner party up on the third floor of the hotel, I was excited when Ron introduced me to another ghost from the past: a redheaded researcher with glasses and a quick smile, Nancy Klimas, formerly an AIDS researcher at the University of Miami. She was one of the early ME/CFS physician-scientists I'd never managed to track down. Her name kept popping up throughout my research as both a collaborator with Komaroff and the first to publish a study linking the disease with abnormal levels of the innate immune cells known as natural killer cells. Whitney, too, had quoted her once on his photography website in the first years after his diagnosis. It's a quote that the advocacy community has used many times, in which she compares her AIDS patients to those with ME/CFS: "I split my clinical time between the two illnesses, and I can tell you if I had to choose between the two illnesses [in 2009] I would rather have H.I.V."

While the others were up at the pasta bar filling their plates and sipping white wine, Nancy and I sat down together at a table, and I asked her to tell me about those early days. She was eager to oblige. She had continued her research ever since publishing her first ME/CFS study while at the University of Miami in the late 1980s. Now she was working at Nova Southeastern University, also in Florida. And, during those years, the patients had continued to seek her out.

"I was a brand-new doctor," she said, laughing, referring to when she was treating patients in the 1980s. "I still had braces on my teeth." For Klimas, the story of ME/CFS revolves around the fact that most of the people who get it are women; the ratio of women to men is as high as four to one, according to the CDC. It falls into that category of so-called contested diseases that primarily affect women, such as most autoimmune diseases. Diseases often dismissed as hysteria. Klimas's first patient was a woman. After that, word quickly got around that she was treating women with ME/CFS, and the numbers grew. She told me a story about the day in 1996, when Hillary Johnson's book was first published, and how her university had instructed her that she was going to have to do an interview about the book on NPR the next day—even though she hadn't even read it yet.

"Luckily Hillary had indexed her book," she said. "I grabbed a copy and looked myself up, and thought, *Phew, OK, at least I'm a good guy.* I wasn't sure before that." The NIH scientist Stephen Straus was one of the bad guys I remembered. She laughed and smiled and adjusted her glasses, then added: "It was a controversial book. It wasn't altogether wrong." Then she grew more serious and paused to consider how the disease had changed her own life.

"I was very interested in women's health, and this was soul satisfying for me," she said. "A woman comes along, and it's dismissed as hysteria. That's truly what happened with this disease. That's what this was all about. It was one of those women's diseases that gets ignored. It turned out to be transformational for me." She paused again, then added, "At times, it wasn't good for my career." I saw the irony in the fact that perhaps this mystery disease, often dismissed as nothing more than "female hysteria," was finally gaining serious attention, in part, as a result of the story of Whitney Dafoe, a handsome young man with a brilliant father who agreed to be its poster child.

Before leaving the conference on the second day, I walked down the hill from the auditorium and returned to the stately building with the white pillars out front and the wide green lawn. I walked up the same steps where FDR spoke and through the hallways, down to Francis Collins's office. When he saw me, I mumbled something about Ron Davis and a book. He nodded his head, smiled, and said I could give him a call anytime, but he was busy at that moment. When I called later, he told me, "The history of this has been very frustrating for anybody who lived through it. Many turned a blind eye. I tell people: don't judge us by what we did ten years ago. Judge us by what we do now."

Chapter 12

The Superman

THE SUN GLISTENING OFF the slick blue ocean flickers through the trees, blinding me. The bright light flashes on and off and on and off through the train windows, as it races past the Pacific coast. My beautiful daughter sits across from me, squinting her eyes at the vast horizon, a book of poetry in her lap, and she's smiling.

I flew down south to Irvine to join Kaily for the train ride home during her summer break from school on the Coast Starlight Express. The train hugs the shoreline from the Los Angeles train station much of the three hundred miles to the agricultural fields of Salinas—a twenty-minute car ride from our home. The tide is far out. A perfect time for a run, I think. Oil rigs stick up oddly from the shimmering waters, which means we must be closing in on Santa Barbara. There's nothing between the train and the ocean but some scrub brush, then sand, then the white foam from the waves. A few surfers. And sky.

I'm bringing my daughter back home, I think, and it catches in my throat. Every time one of my grown children returns home, this happens. I think of Whitney, and a similar tug at my heart makes me long to see Ron bringing his son back home again to him too, but this time healthy and strong. Outside, hawks float effortlessly on the air, and I think about how Whitney should be here to take photographs of all this. Of the hawks and ocean, the old lady staring out at the sea, a young woman seated next to her fiddling with her fishnet stockings, and the family laughing and chatting in Chinese. And some of Stephanie Land's words about Whitney come to mind: "He found beauty in everything—the rise of a burned spruce tree, the surface of a scummy pond. Whitney was a wanderer. A lover of nature, his camera always ready by his side."

After flying home from the conference back East, I'd returned to my cubicle at work and told the office about Ron's upcoming nanoneedle study in the journal *Proceedings of the National Academy of Sciences.* I figured Hanae, the genetics writer, would want to write a press release about it. Part of our job was to promote scientific papers published by Stanford faculty, pushing the news out to the media as quickly as possible.

At my desk, I opened my email, wading through the long list that had grown there while I was gone on my trip. I opened one from the Open Medicine Foundation that featured a video of Ron sitting in his leather chair in his library at home, with Rex, the tortoise, just out of view. Ashley routinely records videos of him from that spot to update the ME/CFS community and other donors on how their money was being put to use.

I scrolled down farther and stopped at another email, this one from Whitney's friend Marta. She wrote that she had been able to track down Whitney's monk, Shenyen. Further down was an email from Shenyen himself. I was excited to tell Whitney, but I knew I'd have to wait awhile, until the next time he took Ativan.

The nanoneedle study, when it hit the media, proved a major success. Both Rahim and Ron were interviewed by reporters from around the globe. Hopes of finally getting a tool into doctors' offices that could definitively diagnose the disease ran high. Our office was flooded with media calls. Headlines announced Stanford had found scientific proof of ME/CFS and discovered the first potential simple blood test for the disease. Even the NIH director, Francis Collins, tweeted about it.

"I'm hoping this will help the medical community accept that this is a real disease," Ron was quoted as saying in a *San Francisco Chronicle* article under the headline "Stanford Discovery Validates Chronic Fatigue Syndrome, Could Improve Diagnosis."

The study used blood samples from people with ME/CFS compared to blood from healthy people. Scientists then exposed the cells to salt, which acts as a stressor on cells and causes them to exert energy to pump out the extra salt. Ron's earlier ME/CFS studies analyzing the citric acid cycle had shown that diseased blood cells could not produce as much energy. The technology the scientists used to test changes from cells' energy expenditure contains thousands of electrodes in a microchip about the size of a corn kernel that measure the electricity passing through the cells. The change in electrical activity is correlated with the health of the cell. The idea is to stress the cells, then compare how each sample affects the flow of the electrical current. A big change is a sign that the cells are in trouble.

When cells from both samples were then passed through the microchip, those from the ME/CFS patients showed a major increase in the expenditure of energy, indicating that the diseased cells had been altered by having to pump out the salt, while samples from the control group remained unchanged. Differentiating between the two groups was now easy.

"It was a very, very clear signal," Ron was quoted as saying in an article in *STAT*.

Rahim, who had by then left Ron's lab and was on faculty at UC Irvine, added, "When they face this new environment, their reaction is different from the reaction of healthy cells." When I emailed Rahim to find out more about the study, he emailed back that, depending on future funding, he hoped to develop a prototype for a simple, handheld ME/CFS diagnostic device to be used in a doctor's office within a year or two. I was excited. Ron was now so close to crossing one of the hurdles he had set up for himself years before: creating a diagnostic test for ME/CFS.

I set up another appointment to meet with Ron at his lab to celebrate the happy news. When the day arrives, I walk across the street from my office during one of my lunch breaks, wave to the security guard in the lobby, smile at the rows of Ron's patents hanging on the wall, and head around the corner to Ron's office.

He is sitting at his desk. Whitney is still smiling down at him from the photo on the wall facing the desk. Ron's head is bent down over his work when I walk in; then he looks up and smiles.

"Congratulations on the publication," I tell him as I sit down in a chair across from the desk. He nods, but I can tell he has already put that news behind him. He's back at work on his latest hypothesis. He's finally been able to fund his researcher—Robert Phair, a biomedical engineer—who had been volunteering in his lab for several years now—and together they are looking further into the metabolic trap hypothesis. They published a paper on it in the open journal *Diagnostics* a few months back.

"It's looking pretty promising right now," Ron says, with a bit more of an uptilt to his voice than usual. I know he's excited about this theory. He believes it could save his son's life, if only it were true. He knows it may not be. As he repeats over and over again to me whenever we discuss science, it's a scientist's job to disprove his own theories. Only after he utterly fails to do that, then maybe there is something to it. He has yet to disprove this theory, but he's trying.

"We still need to run a lot more tests, but the early results are positive," he continues.

"So can you run through the metabolic trap hypothesis for me one more time?" I ask Ron, and his face lights up.

I can see the cogs in his brain begin to whir, and he pulls out a blank notepad and begins to sketch. He is the professor, and I am his student. A complicated interconnected map of molecular pathways appears on the notepad, with arrows pointing to chemicals that lead to reactions occurring inside a human cell that create interconnected hubs of activity. He continues to draw the cellular processes that keep us breathing, that keep us working and playing and living our lives, processes that he believes are broken inside his son's cells.

As the scientist continues to draw, his breathing slows, and he grows calm. When I ask him to spell the names of some of the enzymes he is drawing, he shrugs me off. I should know by now: this brilliant man can't spell. His mind was built to understand how chemistry works. And it's here within the intricate workings of the molecular pathways of life that he finds hope. Science is what saved him from an unhappy childhood. It's what built his career. Science provides the tools necessary to unveil truths from the apparent chaos of matter, and of life itself. Here, written in these pathways, he'll find answers to the questions that plague him. Fixing broken pathways, he still believes, can save Whitney's life.

A few months later, I got a text from Janet, asking once again if I could come to visit Whitney. He had scheduled a date in the summer to take Ativan again, at home this time. We could meet all together in his bedroom. Ashley would be there, and his general practice physician, Dr.

Richard Lee. Whitney was worried about his teeth and his eyesight, and Dr. Lee would check them out. Finally too, I thought, I could tell him about Shenyen's email. It was always gratifying when I could make Whitney smile.

It was a warm summer evening, and the air outside sat soft on bare arms. Inside his bedroom Whitney was enjoying socializing with the three physicians who had gathered around his bed. One was a researcher from Ron's lab who had decided to see if she could figure out what caused Whitney's remarkable reaction to Ativan by analyzing his blood both before and after he had been given the drug. Whitney was smiling broadly at their silly jokes, like, "How many doctors does it take to change a light bulb?" He seemed to be in a great mood, none of the usual angst showing on his face at all.

Over the many months since I'd begun visiting him, it seemed to me that Whitney was getting somewhat better. Occasionally, he would take off his noise-blocking headphones, and I had watched him look at the screen of my iPhone once in awhile. A year or so ago, he couldn't even look into other people's eyes or listen to them talk. Still, he could only do any of this when he took Ativan, and each time he took it, he suffered for weeks after. Still, like this day, he was now even occasionally allowing a few more people into his room when he was on Ativan.

I thought back to one of the first visitors, Jen Brea, a young woman with ME/CFS about Whitney's age, who had joined me during a hospital visit a few months back.

The family had met her during the filming of *Unrest*, Jen's award-winning documentary about ME/CFS. Whitney had risked his health to allow another cameraman into his bedroom back then, before he'd tried Ativan. On the day at the hospital, Whitney had managed to get his mom to find his fifteen-year-old iPod buried somewhere deep in their basement and bring it with them in the ambulance. During the months prior, he must have been creating a playlist in his mind that he wanted Jen, an ME/CFS activist like him, to post for him online to share with other severe patients like him. Jen and I sat by his hospital bed and fiddled with his iPod for at least an hour or two while Whitney gave us pantomimed instructions on how to find each of the six songs he'd chosen hidden there among the thousands of other songs. He waved his fists each time we got one right. When we found one of them, a song by the Flaming Lips called "Waitin' for Superman," he jerked his head, indicating he wanted the playlist named after the song in honor of his dad. Then Jen read the lyrics out loud while Whitney smiled and nodded along.

Tell everybody waitin' for Superman
That they should try to hold on best they can

On that warm summer evening, when I walked into Whitney's bedroom, he was silently laughing along with his group of doctors. I sat down on an empty folding chair in front of his bed and waited for the laughter to stop to tell him about Shenyen's email.

"I heard back again from Marta," I said. As I hoped, he smiled in response. "She found Shenyen. He's on a three-year retreat of silence in Germany. But he gets a few days off every few months when he's allowed to connect with the outside world, and he sent you an email." When I had earlier told his dad about this, Ron had chuckled at the irony of the monk's years of silence dotted with only occasional reprieves, much like Whitney's own years of silence. He loved to hear stories about Whitney's Buddhist beliefs. They reassured him about the long stretches of isolation his son still continued to suffer through. Next I pulled out a printed copy of Shenyen's email from my backpack and began to read it out loud to Whitney.

"My name is Shenyen, the monk that Whitney hung out with briefly in India in 2006. I just heard from Marta about his illness. I remember Whitney as a bit of a dreamer, which to me jelled nicely with his intention to practice photography back in the U.S. after his India trip. He was quiet and thoughtful, but also easygoing and kind. And of course in India we are all living in slow time, free of clutter and a long way from home, at ease with life and open to its hiddenness. We were happy." I paused, looking up at Whitney. He was nodding his head sagely, in agreement. Then I read on:

"I hear he is now expressing his sweeter aspirations which he nurtured in those days—aspirations to help all beings—in the midst of his illness: encouraging people with the same illness who feel like giving up to give their despair to him and let him absorb it. I know he can be creative with anything life throws at (or gently presents to) him. And I'm sure he understands that no one in this world is 'just sick,' no one in this world is 'simply healthy.' All the best to you, his caregivers, and the noble warrior himself." When I looked up again, Whitney was still nodding his head. He looked so much like a monk himself at that moment that I laughed out loud.

Suddenly, the bedroom door opened, and Whitney's head swiveled around to watch Ashley as she stepped inside. She was dressed in a black leotard, and her pregnant belly stuck out big and round. She looked both happy and a bit terrified.

"Bun in the oven!" she said, staring wide-eyed at her brother. Whitney's eyebrows arched high in astonishment. He grinned, rubbed the knuckles of his fists together and pointed at her belly.

"I'm due in about three months, in October around your birthday," she told Whitney. Together they smiled, then tears filled their eyes, and I looked away. I knew Ashley had already told her brother about her pregnancy a few months earlier—the last time he had taken Ativan—but he was just seeing the proof of it for the first time now. When he hadn't been able to attend her wedding three years earlier, she had been both guilt-ridden and heartbroken for moving on with her life without him. But at least, she hoped, he could be part of this pregnancy and the birth of her child. He reached out and took hold of her hand.

Touching is so rare for him now. I get up to leave and join Janet in the kitchen cleaning up from dinner. They need some privacy, I thought.

I'd read a post that Ashley had written on Facebook after telling Whitney she was pregnant some months earlier. Somehow he understood what she was feeling and made it alright. She wrote: "When I got pregnant I knew the first person I wanted to tell after my husband was my brother.... And at seven weeks I was finally able to see him and pantomime to him that I was pregnant. He immediately burst into inconsolable sobs. I sat with him, and cried with him, while he processed what he was missing. When he finally calmed down he looked into my eyes and pantomimed to me, 'thank you for not waiting.' In that moment I knew exactly what he meant.... I knew my brother. It put to rest every fear I had had of living my life without him by my side."

On October 15, Amalia Rose was born, tiny and perfect; Mom was healthy and fine. The birth was twelve days after Whitney's birthday. Parents and grandparents together posted all kinds of photos online. I was in the hallway outside Whitney's bedroom about a month later, when Ashley brought the baby to meet her uncle for the first time. He turned thirty-six that year—another year gone. When Ashley opened his door, Whitney motioned for everyone else in the bedroom to leave. He wanted to be alone with Ashley to meet the baby and greet the new life. An hour or so later, when I returned, Ashley was sitting by his side, the baby cradled in her arms. Whitney, lying on his side, was holding the Buddhist necklace made out of thin green string that always hangs around his neck, pressing it gently against Amalia Rose's tiny foot, his eyes closed in prayer.

Epilogue

DEEP INTO THE CORONAVIRUS pandemic in the spring of 2020, my visits to Whitney stopped. I hunkered down at home, social distancing with my husband from everyone else. Whitney, with his weakened immune system, could no longer have any visitors, nor could Ron, since he is well past the age of sixty-five and has heart problems. I started working full-time at home from my office in my daughter's bedroom and grew obsessed with the hummingbirds that flitted through the trees outside the bedroom window. As always, I prayed for my own children, still far away, to keep healthy and safe.

I wrote about the coronavirus, noting that China was quickly able to map the virus's genome soon after the initial outbreak, which helped speed research along. And I thought proudly of how much Ron's career had helped to make that happen. I wrote stories about ICU heroes, about new research into gastrointestinal symptoms, and potential drug treatments. Over and over again, the tragedy of this new pandemic revealed to me just how reliant the world is on good science to save lives. The scientists themselves, like the NIH's Anthony Fauci, became the new leaders so many of us depended on to steer our nation out of this growing crisis.

I also began to notice an alarming number of media stories about the possibility that this new virus could trigger ME/CFS. Some were saying how a portion of COVID-19 patients had symptoms like muscle aches and severe fatigue that lingered on so long they were calling it post-viral fatigue. I knew that viruses could trigger ME/CFS, so when I got an email from the Open Medicine Foundation announcing it was raising funds to launch Ron's planned study of the connection, I wanted to talk to him. (By now, the OMF had become a fundraising machine, reaching $19 million in donations for ME/CFS research.) The two of us set up a Zoom meeting, the new way everyone was communicating these days. For four years now, he'd been a part of my life, and I was looking forward to the meeting.

Up on the computer screen, Ron smiled a hello, then launched right into the science, as he always does.

"We want to do a long-term study to see if—and if so, how—the COVID-19 converts to ME/CFS," Ron told me. "Technically, to be diagnosed with ME/CFS you have to have fatigue for six months. The virus hasn't been around long enough for that yet. We do know that about 10 percent of those who get mono, which is also a viral disease, come down with ME/CFS. Virus outbreaks have triggered ME/CFS in the past, so it's a big concern. It's occurred to us that the conversion might occur during the active coronavirus infection, so we want to start getting extensive molecular data from patients when they're still sick with the virus and follow them for at least four years. This could enable us to discover the cause of ME/CFS. Until now we've only been able to study patients who have already had the disease for several years. " Ron was clearly excited about doing the study.

We talked on for a bit about other studies continuing at the lab. He elaborated on the latest scientific updates on the metabolic trap research and he said that they recently had a breakthrough. They were able to show that the trap was more than just sophisticated modeling and demonstrated that it actually worked in yeast.

"Now we can look for a drug or a condition that will get yeast out of the trap. This treatment then might work on patients and cure them if the trap is causing ME/CFS," he said. When I asked him about how the family was faring through this pandemic, he smiled that crooked smile of his and said that he and Janet and Whitney were all being very careful to avoid any exposure to COVID-19. Then he grinned and turned to show me just how long his hair had grown. "I've almost got a ponytail by now!"

One day, not long after that, I sat down at my laptop, opened my email, and jerked my head up with surprise. There was a message from an address that I'd never seen before: whitneydafoe. I figured either Janet or Ashley had used Whitney's email to send me a message, but when I opened it, I knew right away it was from Whitney himself.

"Hi Tracie! It's an email from me!" I read in total shock. "I'm experiencing some adrenaline or something that's giving me a brief time of being able to use my phone a little. Not sure why probably one of the experimental drugs I'm always trying."

Over the next few months, Whitney kept writing, sending emails from his cell phone. He speculated that Abilify, a new drug he'd begun taking recently in microdoses of 2 mg a day, may have made a difference. He didn't know how long it would last—short improvements in the past always had disappeared so fast—so he wrote for hours on end, trying his best to communicate as much as he could of his many long-hidden thoughts. He wrote too much, and it sometimes made him sicker, but he kept on writing until he dropped. He'd sit on the toilet near his bed and write. And he'd write from bed. He posted long essays to Facebook about his dad's research, and the stigma of this disease, and how the NIH was still rejecting research grants. (His dad's latest request for nanoneedle funding to develop it for use in clinics as a diagnostic test had been turned down.)

Whitney emailed ex-girlfriends and old friends, finally getting a chance to tell them, in his own voice, how much he loved them and missed them still. For awhile at least, some of his intense loneliness subsided. He struck up a much-longed-for email correspondence with Shenyen the monk, and together they philosophized. "The Buddha referred to old age, sickness and death as 'divine messengers,'" Shenyen wrote. "They come to awaken us from the sleep of an unlived life."

Whitney kept writing to me as well. I asked him many questions, about his worst times and his best, how his faith in his dad kept him going and how Buddhism, too, helped keep him alive. Despite his many years of suffering and all his continuing pain and isolation, Whitney's love for life and desire to help other people with his disease continue on. He plans for the day his father finds a cure and he can go for a drive in his car. He'll garden, listen to music, watch movies, go to Giants baseball games, *play* baseball(!), cook, eat, and go on long hikes. "And strangely," he wrote me, "buy a pair of shoes (because of what it signifies—walking in public outside, wearing clothes, feeling more human)." He hasn't worn a pair of shoes for a long time.

"I often feel like someone hit the pause button on me and fast forward button on the rest of the world," he wrote.

One day, during this correspondence, I got a text from Janet. I'd been reading her many tweets and Facebook posts that alternated between adorable photos of her new granddaughter and repostings of Whitney's long Facebook messages that she decorated with lots of exclamation points. She texted me that Whitney was thinking about taking his first shower in years, a longing he'd written to me about in his emails. A shower, I thought—now wouldn't that be grand.

I thought about my conversation with Laura Hillenbrand, the author of *Unbroken,* whom I had talked to so many years back, before I had met Whitney, when I was hoping to understand what it was like to be sick with ME/CFS. She, too, had been bedridden for years and then had recovered enough to be able to write books from bed. My own hopes soared as I remembered the awe in her voice when she told me that she had recovered even more after that, enough to leave her home and drive thousands of miles across the country to a new home with her boyfriend.

"We drove 4,300 miles," she told me. "Through the Rockies and the Badlands all the way to Oregon." And a vision of Whitney appeared, on another one of his "super epic road trips" driving cross-country with the radio blaring, leaving the Bay Area behind, maybe driving down south to Mexico, or north to Alaska, or wherever it was he traveled in his mind all that time he spent alone in his room with a future still to imagine.

Acknowledgments

Thanks to those in the ME/CFS advocacy community who provided a wealth of resources including in-depth research, and historical archives—that contributed to the writing of this book. For your passion for the truth and persistence in getting the story told.

Tracie White & Ron Davis

From Ron Davis

Thanks to my wife, Janet Dafoe, for her support and help in making this book a reality. We are a team. She both encourages my research and endlessly works on her own in the fight to end ME/CFS. She has remained at our son Whitney's side throughout his illness, providing endless love and care. For this she can never be thanked enough. Donations to my lab's ME/CFS research can be made to the Stanford University Chronic Fatigue Syndrome Research Center at: http://med.stanford.edu/sgtc/donation.html.

From Tracie White

My deepest gratitude to those sick with ME/CFS, including Whitney Dafoe, who told me their stories often at risk to their own health. Among them Gerald and Janice Kennedy, Bruce Train, Rivka Solomon, Allison Dykens, Brian Vastag, Hillary Johnson, Jamie Seltzer, Jen Brea, Mary Schweitzer and the other Tahoe women who spoke with me, and many others I can't name here. To Whitney's family who opened their hearts and home to me, Janet Dafoe and Ashley Haugen and, of course, to my brilliant colleague and friend, Ron Davis.

I'm thankful to so many who helped with the making of this book. My agent Farley Chase for believing in this book and sticking with me through the two long years it took to write the proposal. My editor Krishan Trotman at Hachette who took a chance on a first-time author and helped teach me how to write a book.

My workplace for supporting me. Special thanks to the science writers in the office who had my back including Krista Conger, Bruce Goldman, Erin Digitale, and Hanae Armitage. Thanks to Rosanne Spector, editor of *Stanford Medicine* magazine who first assigned me this story in 2016; Patricia Hannon, Margarita Gallardo, Becky Bach, Mandy Erickson, Alison Peterson, John Sanford, and more. To those talented Santa Cruz writers "the Abbey-ites" whose advice was invaluable. Susan Sherman, Amy Ettinger, Jill Wolfson, Shelly King, and more. To Jonathan

Rabinovitz along with so many other editors from my past who continually fixed my misplaced commas often without complaint. Peggy Townsend, Tom Long, Mike Blaeser and other former co-workers at the *Santa Cruz Sentinel* including photographer Shmuel Thaler.

To the many journalists, researchers, and scientists who helped inform this book including David Bell, David Tuller, Cort Johnson. Bruce Schaar, Laurel Crosby, Katrina Hong and others at the Stanford Genome Technology Center. And Mary Dimmock for her meticulous research and ability to communicate it to me. To Adrianna Baires for always being there at the door to Whitney's hospital room with lunch.

Most of all to my family and friends who supported me including my father Richard White, now passed. My mother Lyn White; my sister, Lisa Pepperdine, who listened intensely to my endless mumblings over the reorganization of chapters; Zak, Max, and Tony Pepperdine; Kellie White and Santiago, Santi and Andres Montufar; Maria Gaura and Mimi Rudd. To Jana and Julia Davids, who constantly encouraged me; novelist and college roommate Mary Smathers and our emergency sessions in Moss Landing. And author Kris Newby, colleague, friend, and role model.

To Michael Ondaatje for writing the novel *The English Patient* about hope and healing which inspired the original title of this book *The Invisible Patient.* Whitney adamantly shot that title down. ME/CFS patients have been called invisible for far too long.

To my loving husband Mark Dorfman, who many years ago said it was OK to be an English major, and to my two children, Kaily and Ben, who opened my heart.

Bibliography

Allan, Nicole. "Who Will Tomorrow's Historians Consider Today's Greatest Inventors?" *Atlantic*. November 2013.

Allday, Erin. "Stanford Discovery Validates Chronic Fatigue Syndrome, Could Improve Diagnosis." *San Francisco Chronicle*. April 29, 2019.

Barte, Barb. "Truckee Teachers Recount 'Malady.'" *Tahoe World*. October 24, 1985.

BBC Newshour. "Scientist Dad Searches for Cure for Sick Son." October 19, 2015. bbc.co.uk/programmes/p035n9g6.

Bekiaris, Angela. "Living with Chronic Fatigue Syndrome." *People*. April 12, 2018.

Bell, David S. *The Doctor's Guide to Chronic Fatigue Syndrome: Understanding, Treating, and Living with CFIDS*. Reading, MA: Addison-Wesley, 1995.

Berg, Paul, David Baltimore, Sydney Brenner, et al. "Summary Statement of the Asilomar Conference on Recombinant DNA Molecules." *Proceedings of the National Academy of Sciences* 72, no. 6 (June 1975): 1981–1984.

Boffey, Philip. "Fatigue 'Virus' Has Experts More Baffled and Skeptical Than Ever." *New York Times*. July 28, 1987.

Botstein, David, Robert L. White, Mark Skolnick, and Ronald W. Davis. "Construction of a Genetic Linkage Map in Man

Using Restriction Fragment Length Polymorphisms." *American Journal of Human Genetics* 32, no. 3 (May 1980): 314–331.

Bowers, Elledge. "Dr. Ron Davis Speaks at SF Demonstration #MillionsMissing." Video. 2016. youtube.com/watch?v=GhIZu3lGnFA.

Bowman, Chris. "Mystery Sickness Hits Tahoe." *Sacramento Bee*. October 11, 1985.

Brea, Jennifer. "Meeting Whitney." Medium. June 3, 2019. medium.com/@jenbrea/meeting-whitney-cf179fdad0a9.

"Bringing Engineers to Medicine with Dr. Ron Davis, Stanford Genome Center." *Mendelspod* (podcast). November 17, 2011. mendelspod.com/podcast/rondavis.

Brody, Jane E. "New Recognition for Chronic Fatigue." *New York Times*. November 27, 2017.

Brody, Jane E. "Personal Health." *New York Times*. October 9, 1996.

Brown, June Gibbs. "Audit of Costs Charged to the Chronic Fatigue Syndrome Program at the Centers for Disease Control and Prevention (A-04-98-04226)." US Department of Health and Human Services. May 10, 1999. oig.hhs.gov/oas/reports/region4/49804226.htm.

Buchwald, Dedra, Paul R. Cheney, Daniel L. Peterson, et al. "A Chronic Illness Characterized by Fatigue, Neurologic and Immunologic Disorders, and Active Human Herpesvirus Type 6 Infection." *Annals of Internal Medicine* 116, no. 2 (January 1992): 103–113.

Chakradhar, Shraddha. "An Experimental Test May Help Confirm Cases of Chronic Fatigue Syndrome." *STAT*. April 30, 2019. statnews.com/2019/04/30/experimental-test-may-help-confirm-chronic-fatigue-syndrome.

"Chronic Fatigue Possibly Related to Epstein-Barr Virus—Nevada." Centers for Disease Control and Prevention. May 30, 1986. cdc.gov/mmwr/preview/mmwrhtml/00000740.htm.

"Chronic Fatigue Syndrome: CDC and NIH Research Activities Are Diverse, but Agency Coordination Is Limited." US General Accounting Office. 1998. gao.gov/products/hehs-00-98.

The Chronic Fatigue Syndrome Research Center (CFSRC). Stanford School of Medicine. med.stanford.edu/sgtc/donation.html.

Cohen, Jon, and Rodrigo Pérez Ortega. "Goodbye Chronic Fatigue Syndrome, Hello SEID." *Science.* February 10, 2015.

Collins, Francis S. *The Language of Life: DNA and the Revolution in Personalized Medicine.* New York: Harper Perennial, 2011.

Collins, Francis, and Walter Koroshetz. "Moving Toward Answers in ME/CFS." National Institutes of Health. US Department of Health and Human Services. September 27, 2017. directorsblog.nih.gov/2017/03/21/moving-toward-answers-in-mecfs.

Davis, Ronald. "A Study of the Base Sequence Arrangements in DNA by Electron Microscopy." PhD diss. California Institute of Technology, 1970. thesis.library.caltech.edu/3675.

Day, W. "Raggedy Ann Syndrome." *Hippocrates.* July 1987.

Dimmock, Mary, and Matthew Lazell-Fairman. "Thirty Years of Disdain (Background): How HHS and a Group of Psychiatrists Buried ME." 2015. dropbox.com/s/bycueauxmh49z4l/Thirty%20Years%20of%20Disdain%20-%20Background.pdf?dl=0.

Donaldson, Sam. "CFS and the CDC's Failure to Respond." *Primetime Live.* ABC-TV. 1996. youtube.com/watch?v=AW0x9_Q8qbo.

Doucleff, Michaeleen. "For People with Chronic Fatigue Syndrome, More Exercise Isn't Better." National Public Radio. October 2, 2017.

Dremann, Sue. "Chronic Fatigue Syndrome Saps Its Victims, but New Research May Find the Cause." Palo Alto Online.

July 10, 2015. paloaltoonline.com/news/2015/07/10/chronic-fatigue-syndrome-saps-its-victims-but-new-research-may-find-the-cause.

Enserink, Martin. "CDC Struggles to Recover from Debacle over Earmark." *Science.* January 7, 2000. science.sciencemag.org/content/287/5450/news-summaries.

"Ensuring You See What Matters Most." Metabolon. metabolon.com. Accessed March 7, 2020.

Esfandyarpour, Rahim, Alex Arron Kashi, Mohsen Nemat-Gorgani, Julie Wilhelmy, and Ron W. Davis. "A Nanoelectronics-Blood-Based Diagnostic Biomarker for Myalgic Encephalomyelitis/Chronic Fatigue Syndrome (ME/CFS)." *Proceedings of the National Academy of Sciences* 116, no. 21 (May 2019): 10250–10257.

Fletcher, Mary Ann, Kevin J. Maher, and Nancy G. Klimas. "Natural Killer Cell Function in Chronic Fatigue Syndrome." *Clinical and Applied Immunology Reviews* 2, no. 2 (April 2002): 129–139.

Fluge, Oystein, Ingrid G. Rekeland, Katarina Lien, et al. "B-Lymphocyte Depletion in Patients with Myalgic Encephalomyelitis/Chronic Fatigue Syndrome: A Randomized, Double-Blind, Placebo-Controlled Trial." *Annals of Internal Medicine* 170, no. 9 (May 2019): 585–593.

The Forgotten Plague. Documentary. Directed by Ryan Prior and Nicole Castillo. USA. 2015. forgottenplague.com.

Fukuda, Keiji, Stephen E. Straus, Ian Hickie, et al. "The Chronic Fatigue Syndrome: A Comprehensive Approach to Its Definition and Study." *Annals of Internal Medicine* 121, no. 12 (December 1994): 953–959.

"Happy 75th Birthday to Scientist Ronald W. Davis, PhD." Open Medicine Foundation. July 12, 2016. youtube.com/watch?v=Tc0tf-Hb_wE.

Henig, Robin Marantz. "Tired All the Time." *Washington Post.* June 30, 1987.

Hillenbrand, Laura. "A Sudden Illness." *New Yorker.* June 30, 2003.

Holmes, Gary P. "A Cluster of Patients with a Chronic Mononucleosis-like Syndrome." *Journal of the American Medical Association* 257, no. 17 (1987): 2297–2302.

Holmes, Gary P., Jonathan E. Kaplan, Nelson M. Gantz, et al. "Chronic Fatigue Syndrome: A Working Case Definition." *Annals of Internal Medicine* 108 (1988): 387–389.

The Human Genome Project. National Human Genome Research Institute. genome.gov/human-genome-project.

"In Memoriam." CFIDS/ME Patient Memorial List. National CFIDS Foundation. 2015. ncf-net.org/memorial.htm.

"Is That What's Wrong with Me?" *20/20.* ABC News. July 31, 1986.

Johnson, Cort. "Dr. Nancy Klimas on Her New Clinic and More." Phoenix Rising ME/CFS Forums. June 17, 2010. forums.phoenixrising.me/threads/dr-nancy-klimas-on-her-new-clinic-and-more.59550.

Johnson, Cort. "Urgency: Ron Davis and His (Non-NIH Funded) ME/CFS Collaborative Research Center." Health Rising. December 22, 2017. healthrising.org/blog/2017/12/21/urgency-ron-davis-chronic-fatigue-research-center.

Johnson, Hillary. "Journey into Fear." *Rolling Stone.* August 13, 1987.

Johnson, Hillary. "Journey into Fear: Part Two." *Rolling Stone.* August 13, 1987.

Johnson, Hillary. "Osler's Web 2.0!" OslersWeb.com. Accessed February 28, 2020.

Johnson, Hillary. *Osler's Web: Inside the Labyrinth of the Chronic Fatigue Syndrome Epidemic.* Lincoln, NE: iUniverse, 2006.

Jones, James F., C. George Ray, Linda L. Minnich, et al. "Evidence for Active Epstein-Barr Virus Infection in Patients with Persistent, Unexplained Illnesses: Elevated Anti-Early Antigen Antibodies." *Annals of Internal Medicine* 102, no. 1 (January 1985): 1–7.

Kashi, Alex A., Ronald W. Davis, and Robert D. Phair. "The IDO Metabolic Trap Hypothesis for the Etiology of ME/CFS." *Diagnostics* 9, no. 3 (July 2019): 82.

Khazan, Olga. "The Tragic Neglect of Chronic Fatigue Syndrome." *Atlantic*. October 15, 2015.

Klimas, Nancy G., Fernando R. Salvato, Robert Morgan, and Mary Ann Fletcher. "Immunologic Abnormalities in Chronic Fatigue Syndrome." *Journal of Clinical Microbiology* 28, no. 6 (June 1990): 1403–1410.

Komaroff, Anthony L. "Advances in Understanding the Pathophysiology of Chronic Fatigue Syndrome." *Journal of the American Medical Association* 322, no. 6 (August 2019): 499–500.

Komaroff, Anthony L., and Tracey A. Cho. "Role of Infection and Neurologic Dysfunction in Chronic Fatigue Syndrome." *Seminars in Neurology* 31, no. 3 (July 2011): 325–337.

Krieger, Lisa M. "Living with COVID-19 When It Won't Go Away." *Mercury News* (San Jose, CA). June 27, 2020.

Land, Stephanie. "The Love of a Thousand Muskoxen: Grieving a Love Lost to Time and Sickness." *Longreads*. October 24, 2016. longreads.com/2016/10/24/the-love-of-a-thousand-muskoxen-grieving-a-love-lost-to-time-and-sickness.

Mariani, Mike. "A Town for People with Chronic-Fatigue Syndrome." *New Yorker*. December 20, 2019.

Maxmen, Amy. "A Reboot for Chronic Fatigue Syndrome Research." *Nature*. January 3, 2018.

Mays, Patricia. "CDC Diverts Chronic Fatigue Funds." Associated Press. July 6, 1999.

McManimen, Stephanie L., Andrew R. Devendorf, Abigail A. Brown, et al. "Mortality in Patients with Myalgic Encephalomyelitis and Chronic Fatigue Syndrome." *Fatigue: Biomedicine, Health & Behavior* 4, no. 4 (2016): 195–207.

Mertz, Janet E., and Ronald W. Davis. "Cleavage of DNA by R1 Restriction Endonuclease Generates Cohesive Ends." *Proceedings of the National Academy of Sciences* 69, no. 11 (November 1972): 3370–3374.

Metabolon. "Ensuring You See What Matters Most." Metabolon .com. Accessed March 7, 2020.

Mukherjee, Siddhartha. *The Gene: An Intimate History.* London: Bodley Head, 2016.

National Academy of Sciences. *Beyond Myalgic Encephalomyelitis /Chronic Fatigue Syndrome: Redefining an Illness.* 2015. iom .nationalacademies.org/Reports/2015/ME-CFS.aspx.

National Institutes of Health. Office of Budget. officeofbudget .od.nih.gov/index.htm.

Open Medicine Foundation. Financial Information. omf.ngo /f-info.

"Paul Berg: Facts." The Nobel Prize. nobelprize.org/prizes /chemistry/1980/berg/facts. Accessed March 7, 2020.

Phoenix Rising: Supporting People with Chronic Fatigue Syndrome (ME/CFS). phoenixrising.me.

Prior, Ryan. "He Pioneered Technology That Fueled the Human Genome Project; Now His Greatest Challenge Is Curing His Own Son." CNN. May 13, 2019. cnn.com/2019/05/12 /health/stanford-geneticist-chronic-fatigue-syndrome -trnd/index.html.

Prior, Ryan. "Viewpoint: Telling the Hidden Story of Chronic Fatigue Syndrome." *USA Today.* June 10, 2013.

Rehmeyer, Julie. *Through the Shadowlands: A Science Writer's Odyssey into an Illness Science Doesn't Understand.* New York: Rodale, 2017.

Rehmeyer, Julie, and David Tuller. "Why Did It Take the CDC So Long to Reverse Course on Debunked Treatments for Chronic Fatigue Syndrome?" *STAT.* September 25, 2017. statnews.com/2017/09/25/chronic-fatigue-syndrome-cdc.

"Ronald Davis." Gruber Foundation. 2011. gruber.yale.edu/genetics/ronald-davis.

"Ronald W. Davis." Stanford Profiles. profiles.stanford.edu/ronald-davis?tab=publications. Accessed February 1, 2020.

Rutherford, Adam. "DNA Double Helix: Discovery That Led to 60 Years of Biological Revolution." *Guardian.* April 25, 2013.

Ryan, Frank. *The Mysterious World of the Human Genome.* Amherst, NY: Prometheus Books, 2016.

"Severely Ill Patient Study." Open Medicine Foundation. November 27, 2019. omf.ngo/2019/11/27/severely-ill-patient-study-study.

Shreeve, James. *The Genome War: How Craig Venter Tried to Capture the Code of Life and Save the World.* New York: Ballantine, 2005.

Snyder, Kim. *I Remember Me* (film). 2000. web.archive.org/web/20071212104112/http://irememberme.com/.

Stanford Magazine. "Tiny Tools for Detangling DNA." September 2000.

Steinbrook, Robert. "160 Victims at Lake Tahoe: Chronic Flu-Like Illness a Medical Mystery Story." *Los Angeles Times.* June 7, 1986.

Stephens, Joe, and Valerie Strauss. "Retaliation Alleged at CDC." *Washington Post.* August 6, 1999.

Straus, Stephen E. "Chronic Fatigue Syndrome." *British Medical Journal* 313 (October 1996): 831.

Straus, Stephen E., Giovanna Tosato, Gary Armstrong, et al. "Persisting Illness and Fatigue in Adults with Evidence of Epstein-Barr Virus Infection." *Annals of Internal Medicine* 102, no. 1 (January 1985): 7–16.

Straus, Stephen E., Janet K. Dale, Martin Tobi, et al. "Acyclovir Treatment of the Chronic Fatigue Syndrome." *New England Journal of Medicine* 319 (December 1988): 1692–1698.

Straus, Stephen E., Janet K. Dale, Ralph Wright, and Dean D. Metcalfe. "Allergy and the Chronic Fatigue Syndrome." *Journal of Allergy and Clinical Immunology* 81 (1988): 791–795.

Strauss, Valerie. "Audit Shows CDC Misled Congress About Funds." *Washington Post.* May 28, 1999.

Tan, Siang Yong, and Yvonne Tatsumura. "Alexander Fleming (1881–1955): Discoverer of Penicillin." *Singapore Medical Journal* 56, no. 7 (July 2015): 366–367.

Tucker, Miriam. "CDC Launches New ME/CFS Guidance for Clinicians." Medscape. July 13, 2018. medscape.com/viewarticle/899316.

Tuller, David. "Chronic Fatigue Syndrome and the CDC: A Long, Tangled Tale." Virology Blog. November 23, 2011. virology.ws/2011/11/23/chronic-fatigue-syndrome-and-the-cdc-a-long-tangled-tale.

Unger, Elizabeth. "Chronic Fatigue Syndrome: It's Real, and We Can Do Better." Medscape. February 25, 2019. medscape.com/viewarticle/908622.

United States Centers for Disease Control and Prevention. "Myalgic Encephalomyelitis/Chronic Fatigue Syndrome: Possible Causes." cdc.gov/me-cfs/about/possible-causes.html. Accessed August 13, 2020.

Unrest. Documentary. Directed by Jennifer Brea. USA. 2017. unrest.film.

Vastag, Brian. "I'm Disabled. Can NIH Spare a Few Dimes?" *Washington Post.* July 20, 2015.

Verghese, Abraham. "A Doctor's Touch." TEDGlobal 2011. July 2011. ted.com/talks/abraham_verghese_a_doctor_s_touch/up-next?language=en.

Wadman, Meredith. "NIH to Double Funding for Chronic Fatigue Syndrome, but Patient Distrust Remains." *Science*. November 10, 2016.

Warlow, Rebecca C. "Rare Footage of FDR at NIH." Circulating Now: From the Historical Collections of the National Library of Medicine. September 10, 2014. circulatingnow.nlm.nih.gov/2014/09/10/rare-footage-of-fdr-at-nih.

Watson, James D. *The Double Helix: A Personal Account of the Discovery of the Structure of DNA*. London: Phoenix, 2011.

Weingarten, Paul. "Mystery Malaise." *Chicago Tribune*. September 3, 2018.

Wessely, Simon. "Chronic Fatigue Syndrome: Summary of a Report of a Joint Committee of Colleges of Physicians, Psychiatrists and General Practitioners." *Journal of the Royal Colleges of Physicians of London*. 1996.

White, Peter D., Michael C. Sharpe, Trudie Chalder, et al. "Protocol for the PACE Trial.'" *BMC Neurology* 7, no. 6 (March 2007).

White, Tracie. "The Puzzle Solver: A Researcher Changes Course to Help His Son." *Stanford Medicine*. Spring 2016.